Further Prai

"Roll over, Kinsey. Mary Roach has done it again. . . . *Bonk* proves that full-bodied research can be riveting."
—*O, The Oprah Magazine*

"For *Bonk*, Mary Roach plunged into the little-known realm of sex research and brought forth an account that is at once revealing—alarmingly so—and very very funny. She studs (forgive me) her journey with a multitude of knee-crossing bits of fact that will enliven bedtime conversation everywhere."
—Erik Larson, author of *The Splendid and the Vile*

"Mary Roach is the funniest writer on sex and death since Sigmund Freud, but without the misogyny and cigars."
—Peter Sagal, host of National Public Radio's *Wait Wait . . . Don't Tell Me!*

"A major triumph. . . . [A]n extended literary comedy routine. . . . *Bonk* reads like a wild night of abandon."
—*Bloomberg*

"Roach ferrets out basic truths and endless absurd details amid mountains of dry science on her chosen subject, in this case, sex. Her most witting collaborators, the scads of sex researchers throughout history, provide plenty of source material. It's a wonderful read, sprinkled with facts you can quote to amaze your friends."

—*San Francisco Chronicle*

"There is science in *Bonk*, lots of it, and it's presented in a most entertaining fashion."

—Jeff Baker, *Oregonian*

"Lush with scientific exploration. . . . *Bonk* is erudite while remaining totally accessible. It is absorbing and quirky, and citations up the proverbial ying-yang give it a credible edge. This is a decidedly satisfying read filled with smarts, stimulating tidbits, and brash wit."

—*Bust*

"I would read Mary Roach on the history of Quonset huts. But Mary Roach on sex? That's a godsend! This book is—if not better than the act itself—then a hilarious and entertaining alternative."

—A. J. Jacobs, author of *The Year of Living Biblically* and *The Puzzler*

"Lucky for us, Mary Roach is a writer impervious to embarrassment. Showing uncommon aplomb in the face of exceedingly odd people and situations, she gamely takes on subjects as weighty as death, the after-life, and now . . . sex. We're all voyeurs at heart, with a desperate need to scratch at these vital, lurid, uncomfortable questions. Thank god we have Roach to do the scratching for us!"

—Hampton Sides, author of *On Desperate Ground*

"Even if there were thousands of science-humor writers, [Roach] would be the sidesplitting favorite. Of course, she chooses good subjects: cadavers in *Stiff* (2003), ghosts in [*Six Feet Over*] (2005), and now a genuinely fertile topic in *Bonk*."

—*Booklist*, starred review

"Roach's ever-present eye and ear for the absurd and her loopy sense of humor make her a delectable guide through this unesteemed scientific outback. . . . Roach's forays offer fascinating evidence of the full range of human weirdness, the nonsense that has often passed for medical science and, more poignantly, the extreme lengths to which people will go to find sexual satisfaction."

—*Publishers Weekly*

"Roach's witty style begets giggling for the right reasons, with well-turned footnotes that are fun to read aloud to prudes."

—*Paste*

"When Mary Roach researches a book, she doesn't do it half way. . . . So it's no surprise that when writing *Bonk: The Curious Coupling of Science and Sex*, she didn't just interview sex researchers, or pore through centuries of lab notes—she actually volunteered to have sex in the name of science. And by that we mean she brought her heroic husband, Ed, from their home in California to an exam room in London to have a physics professor do some real time, 4-D ultrasound footage of their bodies (or at least the relevant parts) in motion. It's difficult to say who deserves more credit for bravery, Mary or Ed,

but the result is a hilarious scene for a book that she calls a tribute to the men and women who brave ridicule to investigate why we do what we do, in bed—or don't."
—*Newsweek Online*

"With *Bonk.* . . . Mary Roach has written a volume so viscerally funny, it's easy to overlook how obsessively she's researched her subject. But Roach's tales of a day with pig inseminators, a hands-on experience with penile implants, and a romp under an ultrasound machine serve as not-so-subtle reminders of her commitment to writing the first-ever comprehensive book on sex research."
—*Seed*

bonk

mary roach

W. W. NORTON & COMPANY
Independent Publishers Since 1923

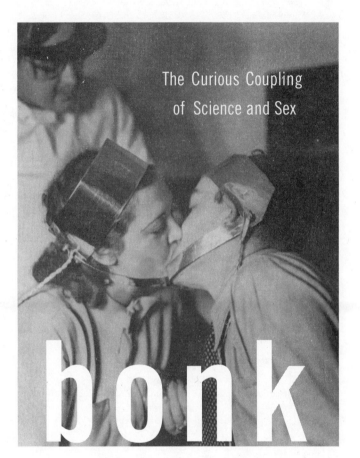

The Curious Coupling
of Science and Sex

bonk

Copyright © 2022, 2008 by Mary Roach

For information about permission to reproduce selections from this book, write to Permissions, W. W. Norton & Company, Inc.
500 Fifth Avenue, New York, NY 10110

For information about special discounts for bulk purchases, please contact W. W. Norton Special Sales at specialsales@wwnorton.com or 800-233-4830

Manufacturing by LSC Harrisonburg
Book design by Abbate Design
Production manager: Anna Oler

Library of Congress Cataloging-in-Publication Data

Roach, Mary.
Bonk : the curious coupling of science and sex / Mary Roach. — 1st ed.
p. cm.
Includes bibliographical references.
ISBN 978-0-393-06464-3 (hardcover)
1. Sex (Biology)—Popular works. I. Title.
QP251.R568 2008
612.6—dc22   2007051990

ISBN 978-1-324-03603-6 pbk.

W. W. Norton & Company, Inc.
500 Fifth Avenue, New York, N.Y. 10110
www.wwnorton.com

W. W. Norton & Company Ltd.
15 Carlisle Street, London W1D 3BS

1 2 3 4 5 6 7 8 9 0

contents

For Woody

foreplay

a man sits in a room, manipulating his kneecaps. It is 1983, on the campus of the University of California, Los Angeles. The man, a study subject, has been told to do this for four minutes, stop, and then resume for a minute more. Then he can put his pants back on, collect his payment, and go home with an entertaining story to tell at suppertime. The study concerns human sexual response. Kneecap manipulation elicits no sexual response, on this planet anyway, and that is why the man is doing it: It's the control activity. (Earlier, the man was told to manipulate the more usual suspect while the researchers measured whatever it was they were measuring.)

I came upon this study while procrastinating in a medical school library some years ago. It had never really occurred to me, before that moment, that sex has been studied in labs, just like sleep or digestion or exfoliation or any other pocket of human physiology. I guess I had known it; I'd just never given it much thought. I'd never thought about what it must be like, the hurdles and the hassles that the researchers faced—raised eyebrows, suspicious wives, gossiping colleagues. Imagine a janitor or a sophomore or the president of UCLA opening the door on the kneecap scene without knocking. Requesting that a study subject twiddle his knees is not immoral or indecent, but it is very hard to explain. And even harder to fund. Who sponsors these studies, I wondered. Who volunteers for them?

It's not surprising that the study of sexual physiology, with a few notable exceptions, did not get rolling in earnest until the 1970s. William Masters and Virginia Johnson said of their field in the late 1950s, ". . . science and scientist continue to be governed by fear—fear of public opinion, . . . fear of religious intolerance, fear of political pressure, and, above all, fear of bigotry and prejudice—as much within as without the professional world." (And then they said, "Oh, what the hell," and built a penis-camera.) The retired British sex physiologist Roy Levin told me that the index of his edition of *Essential Medical Physiology,* a popular textbook in the sixties, had no entry for *penis, vagina, coitus, erection,* or *ejaculation.* Physiology courses skipped orgasm and arousal, as though sex were a secret shame and not an everyday biological event.

One of Levin's earliest projects was to profile the chemical properties of vaginal secretions, the only bodily fluid about which virtually nothing was known. The female moistnesses are the first thing sperm encounter upon touchdown, and so, from a fertility perspective alone, it was an important thing to know. This seemed obvious to him, but not to some of his colleagues in physiology. Levin can recall overhearing a pair of them sniping about him at the urinals during the conference where he presented his paper. The unspoken assumption was that he was somehow deriving an illicit thrill from calculating the ion concentrations of vaginal fluids. That people study sex because they are perverts.

Or, at the very least, because they harbor an unseemly interest in the matter. Which makes some people wary of sex researchers and other people extremely interested. "People invariably draw all these conclusions about me, about why I'm studying this," says researcher Cindy Meston of the University of Texas at Austin. If you are sitting next to Cindy Meston on a plane and you ask her what she does, she will either lie to you or she will say, "I do psychophysiological research." She loses most of them there. "If they persist, I say something like, 'Well, we use various visual and auditory stimuli to look at autonomic nervous system reactivity in various contexts.' That usually does the trick."

Even when a researcher carefully explains a sex-related project—its purpose and its value—people may still suspect he or she is a perv. Last year, I was conversing by e-mail with an acquaintance who was investigating the black market in cadaver parts. She came into possession of a sales list for a company that provides organs and tissues for research. On the list was "vagina with clitoris." She did not believe that there could be a legitimate research purpose for cadaver genitalia. She assumed the researcher had procured the part to have sex with it. I replied that physiologists and people who study sexual dysfunction still have plenty to learn about female arousal and orgasm, and that I could, with little trouble, imagine someone needing such a thing. Besides, I said to this woman, if the guy wanted to have sex with the thing, do you honestly think he'd have bothered with the clitoris?

Early studies of sexual physiology came at it sideways, via studies of fertility, obstetrics and gynecology, and venereal disease. Even working in these areas tended to invite scorn and suspicion. Gynecologist James Platt White was expelled from the American Medical Association in 1851 after inviting medical students to observe a (consenting) woman in labor and delivery. His colleagues had been outraged over the impropriety of a male doctor looking at female genitalia.* In 1875, a gynecologist named Emo Nograth was booed while delivering a talk on venereal disease at the newly formed American Gynecological Society. The sex researcher and historian Vern Bullough, in the 1970s, landed on an FBI list of dangerous Americans for his "subversive activities" (e.g., publishing scholarly papers about prostitution and working for the American Civil Liberties Union to decriminalize, among other things, oral sex and the wearing of dresses by men).

It wasn't until the past half century that lab-based science embraced the pursuit of better, more satisfying sex. Sexual dysfunction had to be medicalized, and the pharmaceutical companies had to get interested. It's still an uphill slog. The current conservative political climate has made funding scarce. Meston plans to seek funding to research fertility—a subject that's easy to fund but does not interest her—simply to help keep her lab afloat. Several researchers told me they keep the titles of their grant proposals intentionally vague, using the word *physiological*, for example, in place of *sexual*.

This book is a tribute to the individuals who dared. Who, to this day, endure ignorance, closed minds, righteousness, and prudery. Their lives are not easy. But their cocktail parties are the best.

. . .

*Incredibly, Victorian physicians practiced gynecology and urology on women *without looking*. Even a catheter insertion would typically be done blind, with the doctor's hands under the sheets and his gaze heading off in some polite middle distance. Fortunately, budding M.D.'s were allowed to look upon—and rehearse upon—cadaver genitals, and that is how they learned to practice the Braille edition of their craft.

eople who write popular books about sex endure a milder if no less inevitable scrutiny. My first book was about human cadavers, and as a result, people assumed that I'm obsessed with death. Now that I have written books about both sex *and* death, God only knows what the word on the street is.

I am obsessed with my research, not by nature but serially: book by book and regardless of topic. All good research—whether for science or for a book—is a form of obsession. And obsession can be awkward. It can be downright embarrassing. I have no doubt that I'm a running joke at the interlibrary loan department of the San Francisco Public Library, where I have requested, over the past two years, papers with titles like "On the Function of Groaning and Hyperventilation During Intercourse" and "An Anal Probe for Monitoring Vascular and Muscular Events During Sexual Response." Last summer, I was in a medical school library xeroxing a journal article called "Vacuum Cleaner Use in Autoerotic Death"* when the paper jammed. I could not bring myself to ask the copy room attendant to help me, but quietly moved over to the adjacent machine and began again.

It's not just library personnel. It's friends and family, and casual acquaintances. It's Frank, the manager of the building where I rent a small office. Frank is a kind and dear man whose build and seeming purity of heart call to mind that enraptured bear in the Charmin commercials. He had stopped by one afternoon to chat about this and that—the Coke machine vandal, odd odors from the beauty school down the hall. At one point in the conversation, I crossed my legs, knocking over a copy of a large hardback that was

---

*They don't mean to tidy up afterward. See p. 208.

propped against the side of my desk. The book slammed flat on the floor, face up. *Atlas of Human Sex Anatomy,* yelled the cover in 90-point type. Frank looked down, and I looked down, and then we went back to talking about the Coke machine. But nothing has been quite the same since.

I like to think that I never completely disappear down the pike. I like to think that I had a lot of miles to go before I got to the point where I was as consumed by the topic as, say, William Masters was. Masters is dead, but I met a St. Louis social worker who used to work in the same building with him. This man told a story about a particularly troubling case he was working on. The father in the case had told him, that morning, that he wasn't all that concerned about his wife gaining custody of their children, because if it happened, he would go and slit their throats. The case was being decided in court the following Monday. The social worker wanted to call the police, but worried that it would be a violation of confidentiality. Distraught, he consulted the only other professional he could find in the building that morning. (It happened to be Thanksgiving.) It was Dr. Masters.

Masters directed the social worker to take a seat on the other side of his enormous rosewood desk, and the man unrolled his dilemma. Masters listened intently, staring at the man from beneath a hedge of chaotic white eyebrows. When the social worker finished talking, there was a moment of quiet. Then Masters spoke: "Have you asked this man whether he has difficulty achieving or maintaining an erection?"

Years ago, I wrote for a women's magazine that tolerated the wanton use of first person among writers such as myself. One month they ran a first-person piece written by a young woman who had had vaginismus. I was acquainted with this woman—I'll call her Ginny—and her piece was tastefully and competently written. Nonetheless, I could not read it without cringing. I did not want to know about Ginny and her boyfriend and their travails with Ginny's clamping vagina.* I would be seeing her at the magazine's holiday party in a few weeks, and now I'd be thinking *clamping vagina, clamping vagina, clamping vagina* as we dipped celery sticks and chatted about our work.

Sex is one of those rare topics wherein the desire for others to keep the nitty-gritty of their experiences private is stronger even than the wish to keep mum on one's own nitty-gritty. I would rather have disclosed to my own mother, in full detail and four-part harmony, the events of a certain summer spent sleeping my way through the backpacker hotels of South America than to have heard her, at the age of seventy-nine, say to me, "Your father had some trouble keeping an erection." (I had it coming: I'd asked about the six-year gap between my brother's birth and mine.) I remember the moment clearly. I felt like Alvy in *Annie Hall*, where he's standing on a Manhattan sidewalk talking to an elderly couple about how they keep the spark in their marriage, and the old man says, "We use a large, vibrating egg."

I've been tripping over the cringe factor all year. It is my habit and preference, as a writer, to go on the scene and report things as they happen. When those things are happening to subjects in sex research labs, this is sometimes impossible. The subjects are queasy about it or the researchers or the university's human subjects review

---

*FYI, it's the newest use for Botox. Because what paralyzes your brow-knitting muscles will just as effectively paralyze your clamping vagina muscles.

board, and sometimes all three. There are times when the only way to gain entrée into the world of laboratory sex is to be the queasy one yourself: to volunteer. These passages make up a tiny sliver of the book, but writing them was a challenge. All the more so for having dragged my husband into it. My solution was to apply the stepdaughter test. I imagined Lily and Phoebe reading these passages, and I tried to write in a way that wouldn't mortify them. Though I've surely failed that test, I remain hopeful that the rest of you won't have reason to cringe.

I promise, no vibrating eggs.

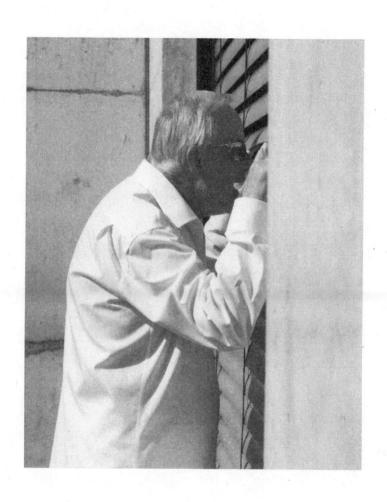

1

The Sausage, the Porcupine, and the Agreeable Mrs. G.

*Highlights from the Pioneers of
Human Sexual Response*

a lbert R. Shadle was the world's foremost expert on the
sexuality of small woodland creatures. If you visit the
library at the Kinsey Institute for Research in Sex, Gender,
and Reproduction, in Bloomington, Indiana, you will find
six reels of audio recordings Shadle made of "skunk and
raccoon copulation and post-coitus behavior reactions."
(Nearby you will also find a 1959 recording of "Sounds
during heterosexual coitus" and a tape of the "masturbatory
sessions" of Subject 127253, which possibly explains why
no one ever gets around to listening to the raccoons.)

Shadle was a biologist at the University of Buffalo in
the 1940s and '50s, back before biology had figured out
most of the basics of life on earth. While today's biologist
spends the days peering through a scanning microscope at
protein receptors or sequencing genomes, the biologist of
the fifties could put some animals in a pen and watch them

have sex. Said Shadle in a 1948 *Journal of Mammalogy* article on the mating habits of porcupines, "Many facts about
these interesting animals await discovery." It was Shadle
who dispelled the myth that porcupines have to have sex
face-to-face; the female protects the male from her spines
by flipping her tail up over her back as a shield.

Here is another fact Shadle discovered by watching
Prickles, Johnnie, Pinkie, Maudie, Nightie, and Old Dad
in the University of Buffalo porcupine enclosure: One
of the males, when sexually aroused, would "rear upon
his hind legs and tail and walk erect towards the female
. . . with his penis fully erected." This was followed by
what Shadle describes as an unusual "urinary shower,"
the particulars of which I'll spare you. Additionally, an
amorous porcupine may hop about "on one front leg and
the hind legs, while he holds the other front paw on his
genitals."

My point is that if you want to understand human
sexual response, then studying animals is probably not the
most productive way to go about it. However, for many
years this was in fact the way scientists—wary of social cen-
sure and career demerits—studied sex. As always, before
science gets its nerve up to try something out on a human
being, it turns first to animals. And it took science a very
long time to get its nerve up to put sexually aroused human
beings under scientific scrutiny. Even the fearless Alfred
Kinsey logged weeks on the road filming animal sex for
study. One particularly productive field trip to Oregon
State Agricultural College yielded 4,000 feet of stag film
featuring cattle, sheep, and rabbits, though no actual stags.
Given the brevity of most animal liaisons, the lessons
learned were rudimentary. Basically, what it came down to
was that, regarding sex, humans are just another mammal.
"Every kind of sexual behavior we had observed or known

about in humans could be found in animals," wrote Kinsey colleague Wardell Pomeroy, who obviously never dropped in to the Yahoo Clown Fetish Group.*

Quite a few scientists in the forties and fifties drove the animal bus way past simple observation and on into the laboratory. I don't want to delve into these experiments because (a) they don't tell us much about people, and (b) they're ghastly. A study that concludes that "removal of the eyes and the olfactory bulbs and deconstruction of the cochlea fails to abolish copulatory responses in the female cat or rabbit" may tell us something about sadism in human beings but not a whole lot about human copulation.

Many people think that the first to dip a toe in the potentially scalding waters of research into human sexual response was William Masters (aided by his associate—and, later, wife—Virginia Johnson). But long before Masters and Johnson and Kinsey became household names, Robert Latou Dickinson was undertaking the unthinkable, in his sunny, cheerfully appointed gynecology practice in Brooklyn Heights, New York. Beginning in 1890, as part of each patient's initial examination, Dickinson would take a detailed sexual history. His patients ran the gamut of turn-of-the-century womanhood; though plenty were well-to-do, he carried a caseload of charity patients as well. Some of these histories were astoundingly intimate.

Subject 177

1897—. . . At 16 . . . slept with another girl—they masturbated each other—suction on her nipples. . . .
Coitus first at 17 and ever since—masturbation was vulvar, vaginal, cervical, mammary. . . . Friction

*Six hundred forty-two members and counting.

against clitoris gives strong pleasure—best is from friction on clitoris to start, then friction against cervix with index finger of other hand. . . . Clitoris not very large but erectile—she has used a clothespin and sausage. . . .

Dickinson writes in the introduction to one of his books that he was inspired and emboldened by "the frank speech" of some of his tenement house patients. Not only were these women at ease talking about their sexuality, but a few eventually allowed him to make observations (with a nurse in the room, always).

Subject 315

1929: Week after period demonstrated climax: legs crossed—her 2 fingers making about inch stroke about 1 to 2 a second—not hard pressure but sway of pelvis and contraction of levator and thigh adduction—rhythmically once in 2 sec or less. Second orgasm, no levator throb—most of desire and feeling outside but "I like inside too."

It might be tempting to dismiss Dickinson as an iconoclastic pervert, but nothing could be further from the truth. He simply believed that lame sex destroyed more marriages than did anything else, and that "considering the inveterate marriage habit of the race," something ought to be done. It was Dickinson who ushered the clitoris into the spotlight. He was an early proponent of the more clitoris-friendly woman-on-top position. Through measurements and interviews he debunked some persistent clitoral myths. For instance, that the bigger ones are more sensitive, and that good girls don't play with them. (Masturbation, he wrote, was "a normal sex experience.")

It was Dickinson's work that inspired Alfred Kinsey to pursue sex research. Kinsey had been, at the time, applying his bottomless research energies to gall wasp speciation. According to Kinsey biographer James Jones, Dickinson— then in his eighties—gave Kinsey his first contacts in the gay male and lesbian communities and turned over dozens of case files of "unorthodox"* patients he'd come across through the years.

Last but, okay, least, we have Dickinson to thank for the innovation of the relaxing picture on the gynecological exam room ceiling. The courtesy was inspired by a grueling afternoon spent staring at the blank ceiling above Dickinson's dentist's chair. I may be dating myself (a turn of phrase that now hits my ears as a euphemism for masturbation), but back in the early eighties, no women's health center was complete without the ceiling poster of a ring of redwood trees shot from below. So ubiquitous was this image that I cannot, to this day, look at a redwood and not feel as though I should scoot down a little lower and relax.

t he first research scientist to make the case for bringing sexual arousal and orgasm into the formal confines of a laboratory was the psychologist John B. Watson. Watson is best known for founding, in 1913, the psychological

---

*For example, the pedophile who dabbled in incest (seventeen relatives, including Grandma) and bestiality. Kinsey's inclusion in *Sexual Behavior in the Human Female* of this man's observations of preadolescent orgasms—and his tacit acceptance of the man's behavior—got him into a pot of hot water that he never really got out of.

movement called behaviorism. It held that human behavior, like animal behavior, was essentially a series of reactions to outside events, an entity easily shaped by reward and punishment. Watson's fame, in no small part, derives from his willingness to study human behavior in a laboratory setting. Most of his subjects were children, most notably Little Albert (no relation to Fat), the eleven-month-old boy in whom he conditioned a fear of white rats. But Watson saw no reason not to bring adults into the lab as well.

Watson chafed at science's reluctance to study human sexuality as it studies human nutrition or planets or porcupine sexuality. "It is admittedly the most important subject in life," he wrote. "It is admittedly the thing that causes the most shipwrecks in the happiness of men and women. And yet our scientific information is so meager. . . . [We should have our questions] answered not by our mothers and grandmothers, not by priests and clergymen in the interest of middle-class mores, nor by general practitioners, not even by Freudians; we . . . want them answered by scientifically trained students of sex. . . ."

Watson's original scientifically trained student of sex may or may not have been Rosalie Rayner, a nineteen-year-old student of his at Johns Hopkins University, with whom he was carrying on an affair. A friend of Watson's, Deke Coleman, says Watson and Rayner "took readings" and "made records" of Rayner's physical responses while they had sex, which would make the pair America's first experimenters (and first subjects) in the laboratory study of human arousal and orgasm. Coleman further claimed that Watson's wife found the notes and data from the experiments, and that these were used as evidence in the ensuing divorce trial.

Watson's biographer Kerry Buckley dismisses the story about the trial as innuendo. Watson was indeed having an affair with Rayner, and the affair did, to use Watson's phrasing, shipwreck his life: When he refused to stop seeing Rayner, he was asked to leave the university and never again managed to work in academia. But Buckley says there is no evidence to support the rumor of the arousal studies making an appearance in the trial. (Mrs. Watson's lawyer did, however, introduce as evidence a cache of love letters, quoted in a different biography of Watson, by David Cohen. Watson expresses his feelings as only the father of behaviorism could do: "My total reactions are positive and towards you. So, likewise, each and every heart reaction.") Buckley is also dubious of the allegation that Rayner and Watson studied their own sexual responses.

Though it would appear that Watson did study *somebody's*. In 1936, a box with John Watson's name on it was discovered in a basement on the Johns Hopkins campus. Inside the box were four scientific instruments. One was a speculum; the other three were a mystery. In the late 1970s, yet another historian, working on a *Journal of Sex Research* article about Watson, heard about the box and contacted its keeper, stating that he wanted to get an expert opinion on the instruments. A photo was taken and mailed to a team of sex researchers in California. "The bent tube with a cage-like end certainly was [an] instrument to insert into the vagina . . . ," began the researchers. I believe them, though I got the sense that an egg beater might have produced the same reply.

The amazing thing about Watson is that, offered a choice between, on the one hand, holding onto respect, prestige, financial security, and tenure at Johns Hopkins and, on the other hand, holding onto the source of his heart reactions,

Watson went with the girl.* Human behavior isn't quite as predictable as the behaviorists made it out to be.

a decade would pass before medical research summoned its courage and hooked up its instruments to live human sex. It was 1932. The researchers, Ernst Boas and Ernst Goldschmidt, knew better than to publish their results in a journal. Their findings appeared quietly on p. 97 of their book *The Heart Rate*. If you are extremely interested in the things that raise or lower a person's heart rate, and exactly how much they raise or lower it, here is a book for you. For example, did you know that "defecating" can briefly bring your heart rate down by eight beats per minute? Or that when a heterosexual man dances with another man—and here I like to picture the two Ernsts in a vigorous foxtrot—his heart rate may rise twenty beats per minute less than it rises when he dances with a woman? The authors include no data on what reading *The Heart Rate* does to one's heart rate, but personal observation puts it solidly between "sitting" and "sleep."

It was Subject No. 69 who agreed to go No. 2 while under cardiac surveillance, and it was also 69 who had sex with her husband, Subject 72, while tethered to the scientists' equipment. Boas and Goldschmidt used a cardiotachometer, which looks from a picture to have been assembled from pieces of Mr. Peabody's Wayback Machine

---

*Watson married Rayner and spent the remainder of his career in advertising. Cohen describes a market research assignment early on in Watson's career at the J. Walter Thompson agency. The mighty John B. Watson was going door-to-door in towns along the Mississippi River, interviewing people about their feelings about rubber boots. Which is not, I suppose, all that far off from a career in psychology.

and the control panel of a B-10 bomber. Subjects wore electrodes held in place by black rubber straps encircling their chests. Boas and Goldschmidt include a photograph of a naked female chest modeling the black rubber harness, lending a glint of illicit eroticism to their otherwise staid endeavor. I'm guessing it's Subject 69's bare bosom on display. Goldschmidt's wife Dora is thanked in the acknowledgments for her contributions to the "experiments that extend over a good part of the day and night," so I'm going to go even further out on a limb and speculate that Subject 69 is Dora and that Subject 72 is hubby Ernst.

Because that's what researchers did back then. Rather than risk being fired or ostracized by explaining their unconventional project to other people and trying to press those other people into service, researchers would simply, quietly, do it themselves.

Whoever the couple was, their heart rates during the encounter ranged from a low of about 80 to a rather shocking 146,* the latter recorded at the third of Subject 69's four orgasms. From the standpoint of sex research, Boas and Goldschmidt's documentation, in 1932, of a woman's

---

*It may comfort you to know that the autopsy data on fatal heart attacks during sex suggest that they are rare. In 1999, a team of German researchers reviewed 21,000 autopsy reports and found only 39 cases. It may or may not comfort you to know that "in most cases sudden death occurred during the sexual act with a prostitute."

Sex researcher Leonard Derogatis cautions that autopsy statistics are misleading. When men die during sex with their mate (as opposed to in a motel room with a stranger), there usually is no reason to do an autopsy. If conjugal sex is taking place, say, three times as often as illicit sex, posits Derogatis in "The Coital Coronary: A Reassessment of the Concept," then those 39 deaths would reflect a truer figure of 156. Derogatis estimates 11,250 sex-related sudden deaths in the United States each year, putting it on a par with hepatitis C, brain cancer, and food poisoning.

multiple orgasms is of far more interest than the rather obvious fact that one's heart beats a lot faster during sex. Alfred Kinsey's data on the prevalence of multiple orgasms, revealed twenty years hence, was met with skepticism on the part of certain segments of the populace who were still adjusting to the notion that women were orgasmic at all. In part, this has to do with the social conservatism of the era. The twenties and thirties were a much looser time than the forties and fifties. I came across a 1950 journal article in which a team of researchers, G. Klumbies and H. Kleinsorge, had recruited a woman who could bring herself to orgasm five times in quick succession. But the authors weren't studying the phenomenon of multiple orgasms; it was a simple study of blood pressure during orgasm. The subject—"our hypersexual woman," as the researchers called her—had been recruited, it would appear, simply for the efficiency and productivity of her orgasmic output. And because she could do it hands-free. (She was using fantasy.) The team had found a way to do its study without recruiting people to have sex at the lab (a risky undertaking in the fifties) or appearing to condone masturbation. "Development and subsidence of the orgasm reflex took place without any physical interference," Klumbies points out in the very first paragraph. In other words, *it's okay—she didn't touch herself.*

Another way to get around the seeming impropriety of laboratory fornication was to so thoroughly bedeck your participants in the trappings of science that what they were doing no longer looked like sex. As was the case in R. G. Bartlett, Jr.'s 1956 study "Physiologic Responses During Coitus." Picture a bed in a small "experimental room." On the bed are a man and a woman. They are making the familiar movements made by millions of other couples on a bed that night, yet they look nothing like these couples. They have EKG wires leading from their thighs and arms, like a pair of lustful marionettes who managed to escape the puppet show and check into a cheap motel. Their mouths are covered by snorkel-type mouthpieces with valves. Trailing from each mouthpiece is a length of flexible tubing that runs through the wall to the room next door, where Bartlett is measuring their breathing rate. To ensure that they don't breath through their noses, the noses have been "lightly clamped." On either side of the bed are buttons for the pair to press, signaling "intromission, orgasm, and withdrawal." When I first read this I pictured an ATM keypad, with different buttons for each event. Then I realized it was simply one button, which I imagined as being attached to a buzzer, providing a madcap game show air, as though at any moment a disembodied voice might ask them, for $500, to name Millard Fillmore's vice presidential running mate.*

I understand why Bartlett did not include photographs in his *Journal of Applied Physiology* article, but I have not forgiven him.

none of the trappings of the Respectable Scientific Endeavor were on hand during the project Alfred Kinsey referred to as "Physiological Studies of Sexual Arousal and Sexual Orgasm." No one was hooked up or plugged into anything other than his or her partner. The studies took place on a mattress laid out on the pine floor of Kinsey's attic in Bloomington, Indiana.

---

*A trick question! Fillmore had no vice president and he never ran for office. He came into power when Zachary Taylor died, and failed, despite repeated efforts, to win a second term. Random quotes suggest his oratory skills might have been the problem. Fillmore's last words (upon tasting a soup): "The nourishment is palatable."

Kinsey is of course best known for his daring, encyclopedic surveys of sexual behavior. (In the 1940s and early '50s, Kinsey—with colleagues Wardell Pomeroy, Clyde Martin, and Paul Gebhard—interviewed 18,000 Americans about their sex lives and published his findings in two ground-breaking, best-selling, ultimately career-tanking volumes.) But Kinsey, a biologist by training, was interested in the physiology of sex, not just the habits of its practitioners. In 1949, Kinsey had plans to set up a dedicated experimental laboratory as part of the Kinsey Institute when it moved to a larger building. He wanted, in essence, to do what Masters and Johnson would do ten years down the road: observe, document, and understand the responses of the human body to sexual stimulation.

The lab never materialized. Kinsey must have sensed it was too risky to go public with such an undertaking. So he went ahead in secret. Thirty couplings—some straight and some gay male—and a similar number of "masturbatory sessions" were observed and filmed in the attic of Kinsey's house. Kinsey had hired a commercial photographer named Bill Dellenback, whose pay, not entirely fraudulently, came out of the institute's budget for "mammalian behavior studies."

Because the work was done in secret, Kinsey didn't recruit his subjects from the public at large. Outsiders— including, says Pomeroy, several "eminent scientists" who had visited the Institute—were filmed if they volunteered and if it was felt they could be trusted, but for the most part, it was an inside project. Kinsey wanted Dellenback to film his own staff. There are three ways to read that sentence, all of them true. Dellenback filmed Pomeroy and Gebhard having sex with their wives and sometimes other people's wives, and he filmed them masturbating. He filmed Kinsey himself masturbating, in one instance, by pushing a swizzle stick* up his staff. Dellenback himself, says Kinsey's other biographer James Jones, reluctantly agreed to masturbate at an attic gathering, though he drew the line at filming himself.

It is difficult to read about the attic sessions and not suspect that there was at least an undercurrent of something beyond research going on up there. Jones describes Kinsey as a voyeur. But the passage Jones uses to illustrate Kinsey's voyeurism, to me, makes an equally strong case that he was simply a biologist studying sex as obsessively as he had studied gall wasps.

---

*More famously, Kinsey employed a toothbrush (bristle end first) for this purpose. This, among other things, caused Jones to describe Kinsey as a masochist, driven by the demons of his repressive upbringing. A past director of the Kinsey Institute told Kinsey's other biographer, Jonathan Gathorne-Hardy, that he viewed the urethral insertions simply as an idiosyncratic form of self-stimulation and that everything else was conjecture. Gathorne-Hardy was invited to the Kinsey Institute fiftieth-anniversary bash, and Jones was not.

A toothbrush, by the way, is alarming but not all that unusual. *Urological Oddities*, a 1948 compendium of memorable cases, includes an "elderly fellow" with a corsage pin that got away from him, a man who died from infection after inserting a twig from the family Christmas tree, and a farmer who "lost a rat's tail." There is always an explanation. The man toting three sets of three-inch surgical steel forceps, for example, insisted that Nos. 2 and 3 had gone in in an effort to remove Nos. 1 and 2, a story that collapsed upon examination, when all three turned out to be in there handle-first. As embarrassing as these hospital visits must have been, they pale in comparison to the Houston man who was taken away, on his back in an ambulance, with a large water tank from a public commode stuck on his penis. "The patient had attempted intercourse with the water-tank hole," reports B. H. Bayer, M.D., in one of those rare, shining moments when urology approaches high comedy.

. . . Kinsey was virtually on top of the action, his head only inches removed from the couple's genitals. . . . Above the groans and moans, Kinsey could be heard chattering away, pointing out various signs of sexual arousal as the couple progressed through the different stages of intercourse. In [his colleague] Beach's estimation, no observer had a keener eye for detail. Nothing escaped Kinsey's notice—not the subtle changes in the breast's skin tone that accompanied tumescence during arousal, not the involuntary twitch of the muscles in the anus upon orgasm—Kinsey saw everything. At one level, it was all very analytic and detached. As Beach looked closer, however, he was certain he detected a gleam of desire in Kinsey's eyes, a look that grew more intense as the action built to a climax.

I wanted to see the purported gleam and decide for myself. I wanted to watch one of the films. Did Kinsey look like a scientist doing research, or did he look like a Peeping Tom? Was he taking measurements? Jotting down notes? Perhaps I could see those too. I contacted Shawn Wilson, the likable gatekeeper for the Kinsey Institute's library and special collections. He replied that if—as Wardell Pomeroy states in his book—notes were taken and data compiled from the sessions in the attic, the institute did not have them. The films themselves, he said, were "not available"—meaning, I think, that they still exist but very few people, and certainly not Mary Roach, get to watch them. In his email, Wilson referred to the footage as "the Kinsey stag film," a fitting enough description, but not one that contributed greatly to its status as scientific documentation.

Kinsey didn't publish research papers about what he

learned from watching his colleagues, but he did include it in his second sex volume, in the chapter "Physiology of Sexual Response and Orgasm." There can be no doubt, in reading this chapter, that Kinsey had a biologist's eye trained upon the proceedings. A Peeping Tom might have noticed that "the anal sphincter may rhythmically open and close" during orgasm, but only a biologist would have noted that people's earlobes swell when they're aroused, or that "the membranes which line the nostrils may secrete more than their usual amounts of mucus." Who but a biologist would have documented the activity of the salivary glands with the approach of orgasm? "If one's mouth is open when there is a sudden upsurge of erotic stimulation and response," Kinsey writes, "saliva may be spurted some distance out of the mouth."

Kinsey didn't supply the average distance covered by the saliva, but I wouldn't be surprised if he'd calculated it. Some years earlier, he measured the average distance traveled by ejaculated semen. Three hundred men, recruited by a well-connected male sex worker, were paid to masturbate on film in the home of an acquaintance of Kinsey's in New York City. Physicians at the time were claiming that "the force with which the semen is thrown against the cervix," quoting Kinsey, was a factor in fertility. Kinsey thought it was bunk, that semen was rarely spurted, squirted, or "thrown," that it mostly just slopped onto whatever surface was closest. In three-quarters of the men, as Kinsey anticipated, that is what it did. In the remainder, the semen was launched anywhere from a matter of inches to a foot or two away. (The record holder landed just shy of the eight-foot mark.) Not one but two sheets were laid down to protect the Oriental carpets.

Kinsey's original plan had been to film *2,000* men

ejaculating. It would be easy to think that Kinsey—who was enthusiastically, though not publically, bisexual—was bringing all these men in because he enjoyed watching them. But if you knew much of anything about Alfred Kinsey, you might, alternatively, take this as an example of the famous Kinsey overkill. In all, the Kinsey team interviewed 18,000 Americans about their sex lives, but Kinsey's hope had been to keep going until they'd talked to 100,000. In his gall wasp days, Kinsey traveled 32,000 miles and collected 51,000 specimens.

With sex research, unlike, say, engineering or genome research, almost everything a scientist does can appear—to the uninformed or close-minded outsider—to be motivated by a perverse fascination with the subject. When, in fact, there's a clear logic to these things. That Kinsey filmed gay male sex workers and their pals for his ejaculation/fertility study could be viewed as a reflection of his own sexual fixations or it could be viewed as simply the most expeditious approach. If you needed three hundred men willing to perform sexually in exchange for quick cash, in 1948, whom would you turn to? In his chapter about the attic sessions, Pomeroy explains that Kinsey's team simply "found it easier to obtain the consent of homosexual couples." (By "homosexuals," he means men. "We were unable to obtain any lesbians," Pomeroy says, as though perhaps they hadn't been in season, or his paperwork wasn't in order.)

Kinsey is admittedly a bit of an extreme case, and it is easy to understand the suspicion that he was perhaps at the very least, as Jones put it, mixing business with pleasure. Even Kinsey's colleague Clyde Martin, now eighty-eight, was uncomfortable with the attic project. Martin refused to be filmed having sex with his wife—or anyone else. "I was not in favor of that," he told me. "I was not part of that. I was married at the time, and I had a wife I loved very much."*

On the other side of the mattress is Wardell Pomeroy, who was adamant about the scientific purity of the project. "The layman can scarcely imagine viewing a sexual scene without having feelings either of stimulation or of disgust, depending on the state of his inhibitions," he wrote in *Dr. Kinsey and the Institute for Sex Research*. "We experienced neither emotion. . . . Speaking for myself I cannot recall a single instance of sexual arousal on my part when I was observing sex behavior, and I am certain this was equally true of Kinsey. . . ."

To be fair to Kinsey, it should be pointed out that gay men weren't the only special-interest group he recruited. Stutterers, amputees, paraplegics, even those with cerebral palsy were observed. Kinsey wanted to document the full spectrum of human sexuality, but it was more than that. He believed these people might have things to teach us about the physiology of sex. And he was right. These groups alerted Kinsey—and the scientific community as a whole—to the complicated and crucial role of the central

---

*Martin's brief and somewhat reluctant affair with Kinsey was made public in both the recent Kinsey biographies. If you bring up gay male sexual orientation, he will quickly change the subject but appears to bear no grudge against his former boss. "I must say, working with a man like Kinsey is a tremendous stimulation," he told me, not choosing his nouns as carefully as he might.

Martin made use of his unique talents, going on to do interviews of his own at Johns Hopkins University. But while Kinsey used his data to promote tolerance and expand notions of sexual "normalcy," Martin pulled a 180, looking for links between promiscuity and disease. His work helped uncover the link between sexually transmitted disease and cervical cancer.

nervous system in sex and reproduction. Kinsey had noted that a stutterer in the throes of sexual abandon may temporarily lose his stutter. Similarly, the phantom limb pain some amputees feel temporarily disappears. Even the muscle spasticity of cerebral palsy may be briefly quieted. The body's limiting factors seem to get shut off.* The organism is driven toward nature's singular goal—conception, the passing on of one's genes—and anything that stands in the way is pushed into the background. Sensory distractions become imperceptible: noises go unheeded and peripheral vision all but disappears—a fact some sex workers use to their advantage, working with "creepers" who emerge from the shadows when the action heats up and go through the john's pockets as easily as if he were unconscious.

The most dramatic example of this biological priority shift is a sexually mediated disregard for pain and physical discomfort. Whatever ails you pretty much stops ailing you during really hot sex. Fevers and muscle aches, Kinsey claimed, briefly abate. Temperature extremes go unnoticed, which must have been a relief for the couples in Kinsey's attic, as it was, depending on the season, either very hot or very cold up there. Handily, the gag reflex is eliminated, even "among individuals who are quite prone to gag when objects are placed deep in their mouths." (Objects! Har.)

To explore the limits of this phenomenon, Kinsey

---

*Non-disabled people, as well, Kinsey observed, enjoy expanded physical prowess under the influence of sexual arousal: "The doubling of the body which is necessary in self-fellation . . . may become possible for some males as they approach orgasm." Or, according to a 2001 *Hustler* article, as they master the yoga pose "the plow" (on one's back, legs flipped up and over the head). Further tips can be gleaned by renting *Blown Alone* or other videos starring superlimber porn star Al Eingang. Wikipedia says that the god Horus was said to engage in autofellatio "every night because ingesting his own semen kept the stars in their places." Only gods get away with excuses like that.

observed and filmed sadomasochistic sex. Which makes sense, but at the same time leaves the reader just a tad bit queasy. Kinsey's "experimental data" indicated that arousal can render a person "increasingly insensitive to tactile stimulation and even to sharp blows and severe injury." If there are cuts, he says, they bleed less. In his discussion of temperature extremes, cigarette burns make a cameo appearance. He is occasionally coy about the source of these injuries, but more often he is baldly straightforward: "The recipient in flagellation or other types of sadomasochistic behavior may receive extreme punishment without being aware that he is being subjected to more than mild tactile stimulation"—surely a source of comfort to anyone who read the toothbrush footnote on p. 33.

It was 1954 when William Masters embarked on his own investigation of sexual physiology. Kinsey was under fire from conservatives. The Rockefeller Foundation, partly because of its funding of Kinsey's work, was the subject of a congressional investigation. (As a result, the foundation pulled Kinsey's funding. He died less than two years later.)

Given the political climate, it was exceedingly brave of Masters—then a gynecologist at Washington University in St. Louis—to undertake such a project. This was to be a large (nearly 700 participants), nonclandestine observational study of human sexual arousal and orgasm. To try to get funding and permission for such a venture in 1954 must have been, well, like trying to do it in 2007. Understandably, Masters went to great lengths to appear as scientific, objective, and morally upstanding as he could. His hiring of a female associate, Virginia Johnson, helped ward off accusations of impropriety (though she was mainly brought on board, Masters said, as a sort of "interpreter" to help him understand a woman's subjective experience of sex). Where Kinsey had actively sought out people on the fringes of American sexuality, Masters made a point of screening out "all individuals with sociosexual aberrancy." (The team observed gay men and lesbians in the lab but did not include them in the sample for this project. More on them later.) The 276 couples who came to his lab were heterosexual, and they were married. Most of them worked or taught at the university. The work was done under the auspices of the Reproductive Biology Research Foundation—no mention of sex—and it was done in a laboratory setting, amid scientific instruments and professionals in white lab coats.

Yet, at the core of it, you had couples fornicating on film. You had women and men masturbating in front of other men and women. You had a man scrutinizing— whether in person or by watching footage—the genitalia of women having orgasms. Moreover, you had sex workers serving as your beta test. Masters and Johnson interviewed 145 qualified individuals and from them chose the cream of the crop—eight women and three men with "obvious intelligence, diverse experience in prostitution, ability to vocalize effectively, and . . . a high degree of cooperation"—to come down to the lab and help the team hone their investigative techniques. (Kinsey had avoided using female sex workers for his observational studies, because, he said, they readily and convincingly faked orgasms.* Masters didn't have to worry about his sex workers faking

---

*How did he know this? It's not what you think. He and a colleague would on occasion hide—with the women's permission—in brothel bedrooms, jotting down observations. At least I think they hid. It's possible they drilled a hole in the wall or rigged up something more high-tech, but I enjoy picturing the two of them peering from behind a set of lurid velour draperies.

it. His subjects were rigged up to a machine that measured heart rate and blood pressure—essentially a lie detector. Heart rate and blood pressure, it turns out, are more reliable indicators of orgasm than they are of deceit.)

Masters and Johnson launched their book-length write-up of the project, *Human Sexual Response,* in 1966. (Medical journals had rejected the team's papers, deeming them pornographic.) "The hate mail was unbelievable," Masters recalled during a talk at the 1983 meeting of the Society for the Scientific Study of Sex. "For the next year and a half, we had extra secretaries . . . just answering mail. . . ."

Eventually, the rancor cooled, and the book went on to become an enduring bestseller and a classic in the field. It is hard to say which contributed most to its acceptance: the cloak of formal science that Masters so assiduously pinned to his work, or the simple fact that times had changed. Nineteen sixty-six was worlds away from 1954.

Unfortunately, the cloak of science was pinned so tight that the book kind of suffocates. A couple under observation is a "reacting unit." An orgasm is rarely just an orgasm; it's "orgasmic phase expression" or an "orgasmic release of sexual tension." A woman who has one during half her sexual encounters is experiencing "a 50 percent orgasmic return." Porn is "stimulative literature," and not getting it up is a "failure of erective performance."

If you can machete through the lingo and the obfuscated writing, you will find an extraordinary body of work. Kinsey—and everyone else who came before—missed a variety of extraordinary things going on between a woman's legs. Take, for instance, the outer labia. Overlooked and ignored, they were thought of simply as packaging. Kinsey was dismissive of the labia majora's role in the sexual chain of events, saying there was no evidence that they "contribute in any important way." Masters and Johnson noticed that, in fact, they do contribute. While other parts swell and even protrude during arousal—because of the extra blood in the tissues—the outer labia thin and flatten. They also, he observed, pull away from the "vaginal outlet."* Masters speculated, in characteristic multisyllabic manner, that this might be "an involuntary neurophysiologic attempt to remove any exterior impediment to the anticipated mounting process." They're making way for the big guy.

No one had expected this, possibly because their inner cousins expand so much. The labia minora enlarge by two or three times their normal diameter. They also, as both Masters and Dickinson observed, change color, turning pink, bright red, or, occasionally, in women who've given birth, a deep wine color. In all of the 7,500 female sexual response cycles that Masters and Johnson watched, no woman who had an orgasm failed to display this "florid coloration" just beforehand. If a sexual partner wants to know whether a woman is faking her orgasms the individual could, barring some logistical hurdles, look for this "sex-skin reaction." Which, by the way, is not to be confused with the "sex flush" (red blotches that may appear on a woman's chest when she's aroused). And the "sex flush," in turn, has nothing to do with the "urge to void during or immediately after intercourse."

Here's something else no one but Masters had noticed. The clitoris hides at a certain point in the proceedings. In the stage of arousal just before orgasm, the visible portion of the clitoris retracts under its tiny foreskin. It disappears from view, potentially creating great confusion and consternation on the part of the person doing the stimulating.

---

*You need a floor plan to keep track of the vaginas in *Human Sexual Response*. There are vaginal floors, vestibules, platforms, barrels, and outlets. Are people having sex, or are they just visiting Crate and Barrel?

Masters points out that the clitoris, at this point, is likely to be too sensitive for direct contact anyway. The shroud of academia pulls away like a foreskin, revealing the readable writer within: "In direct manipulation of the clitoris there is a narrow margin between stimulation and irritation."

Masters and Johnson provide a similar service for men. In the penis chapter,* they describe what they called "postejaculatory glans sensitivity." For many men, once they've ejaculated, continued thrusting on behalf of their partner is chivalrous but exceptionally uncomfortable. The solution to the oversensitive glans scenario, be it penile or clitoral: "Vocalization." Speak up. Throughout *Human Sexual Response*, the researchers encourage open and straightforward communication between partners. It comes as no surprise that they moved on to sex therapy (giving, not getting) following the eleven-year physiology project. Their therapy techniques and writings—as well as the hundreds of therapists they inspired—are the answer to every person who questions the point of Masters and Johnson's lab work. It is hard to overestimate the value of a simple anatomical explanation for a frustrated couple's complaint. Imagine a woman who's been harboring resentment toward her husband for pulling out as soon he's done (and she isn't). Were she to learn that her man is not so much *in*sensitive as *over*sensitive, her resentment would diffuse or, at least, hang its hat on something other than his penis.

Here's another example. Masters and Johnson discovered all manner of physical changes going on in women's vaginas when they're aroused. "Advanced excitement" prompts a portion of the vagina to expand. One theorized purpose is to create an "anatomic basin" to hold the semen

---

*"The Penis."

near the opening to the uterus and thereby up the odds of conception. But the expansion can have an unwanted side effect: "The overdistended excitement-phase vagina gives many women the sensation that the fully erect penis is 'lost in the vagina.'" Sometimes to the extent that the woman mistakenly thinks it's gone limp.

Some of you may be wondering—and some of you really, really may not be—how Masters made his pioneering vaginal discoveries. Two answers. Sometimes subjects were asked to masturbate with an open speculum in place—as Dickinson had had women do years before—while the researcher peered intently up their midline. But Masters didn't want to limit his findings to the arousal and orgasms of masturbation. He wanted to know what was going on with the cervix and the vagina during a typical round of bonk. Obviously, there are logistical problems here. You can't see the hangar when the airship's in the building. William Masters needed a penis that could see.

And so he had one built.

2

Dating the Penis-Camera

Can a Woman Find Happiness with a Machine?

et me state it simply. Women came into Masters and Johnson's laboratory and had sex with a thrusting mechanical penis-camera that filmed—*from the inside*—their physical responses to it. The team shot footage of—as they put it, making arousal and orgasm sound like washing machine functions—"hundreds of complete cycles of sexual response." The dildo-camera* unmasked, among

---

*You can't buy a penis-camera like M & J used, but you can buy a Personal Pelvic Viewer. The PPV is described in the patent as an insertable video camera that enables a "lone female in a room" to watch real-time images of her cervix or her sexual responses on a TV or computer screen. The phrase "lone female in a room" or "lone female at home" is repeated nineteen times, lending a melancholy cast to the typical technicalities of a patent paper. The PPV, at the time I inquired, was being sold by an organization called School of One—no doubt the alma mater of the lone female.

many other things, the source of vaginal lubrication: not glandular secretions but plasma (the clear broth in which blood cells float) seeping through capillary walls in the vagina. It tackled the sucking cervix debate and uncovered the bizarre phenomenon of vaginal tenting (both of which we'll get to later).

This, to me, is as good as science gets: a mildly outrageous, terrifically courageous, seemingly efficacious display of creative problem-solving, fueled by a bullheaded dedication to amassing facts and dispelling myths in a long-neglected area of human physiology. Kudos to the pair of them.

But I have a question. Who *were* these women having orgasms from nothing more than the straight-on, in-and-out motions of a plastic phallus? Some 70 percent of women report that intercourse—ungarnished by any add-on clitoral stimulation—reliably fails to take them all the way to the spin cycle. Remove foreplay and love and lust from the equation, and the orgasms of Masters and Johnson's "artificial coition" subjects are a rather startling achievement.

Especially if you buy what Masters and Johnson had to say about vaginal orgasm: i.e., that it doesn't exist. The team compared the physiological elements of orgasms from (clitorally) masturbating and from intercourse, and they concluded that all the orgasms were, physiologically, the same. And all of them, they maintained, owed their existence to the clitoris. Yet I'm guessing that the artificial-coition machine, because you could not straddle it or position it just so atop you, maintained a frosty distance from the clitoris. (Subjects were either on their backs or on all fours, doggy-style.) So what, then, was bringing these women to their peak? "Penile traction on the labia minora," said Masters and Johnson: Penis going into vagina pulls on labia, which in turn pull on clitoris.

In 1984, a team of Colombian researchers cast doubt

on the notion of labial traction as an instigator of female orgasm. Heli Alzate was a physician and professor of sexology at Caldas University School of Medicine, and Maria Ladi Londoño, his coauthor, was a psychotherapist with a diploma in psychology.* The team brought sixteen sex workers (paid $16 each—several times the going rate of a Colombian trick)† and thirty-two unpaid feminists into their lab to map the erotic sensitivity of the vagina. Like this:

> The examiner, with his or her hands washed, inserted his or her lubricated index and/or middle fingers in the subject's vagina and proceeded to systematically friction both vaginal walls, applying a moderate-to-strong rhythmic pressure at an angle to the wall, going from lower to upper half of the vagina.

When Alzate or Londoño located a subject's sweet spot—which for most was on the front wall, but for some, the lower back wall—the spot was simultaneously pressed and stroked (a maneuver I have seen elsewhere described as a "come here" motion). More than three-quarters of the sex workers Alzate "frictioned" in this manner had a vaginal orgasm. (Londoño brought no subjects to climax; the women said that this was because she wasn't pressing as hard as Alzate.) Only four of the feminists, though aroused, reached orgasm. Perhaps they were feeling uncomfortable

---

*Unfortunately for Londoño, this meant that while Alzate has an M.D. after his name, she must appear in print as a DipPsy.

†Which, in 1976, was $2.50. I learned this from another of Alzate's papers, "Brothel Prostitution in Colombia," this one researched via interviews and "by means of participant observation" (!).

with what many feminists might perceive to be an exploitative scenario.* Or perhaps they were simply less accustomed to sexual encounters with strangers.

Alzate was creating vaginal orgasms, but you couldn't use a penis-camera to bring them on. A male organ in the missionary position travels parallel to the vaginal walls, not at an angle. To prove that their subjects' orgasms were not being caused by traction created by the thrusting motions of the researcher's fingers, Alzate and Londoño set up a separate "simulated intercourse" test. This time no one came.

> Six paid subjects that easily reached climax by stimulation of their vaginal erogenous zones were examined. The examiner rhythmically stimulated the lower third of the vagina with his index and middle fingers, mimicking the movements of the penis during coitus, and for the time required to elicit an orgasm by stimulation of the vaginal erogenous zone. Although a clear traction on the labia minora was evident, all subjects felt only a slight to moderate erotic sensation.

Penile thrusting on its own—with no foreplay or during-play—is, concluded Alzate, "an inefficient method of inducing female orgasm."

Yet Masters and Johnson's artificial coition subjects were getting off on nothing but the thrust. How so?

---

*Alzate insisted these encounters were "ethically acceptable as long as the examiner keeps from being erotically involved with the subject." Only in the mutant universe of sexology could a man with his fingers in a woman who is exhibiting "hyperventilation, . . . rhythmic pelvic movements, vocalizations, and perspiration" not be considered erotically involved.

Were they turned on by the idea of sex with a machine? Is there something about mechanical sex that I'm failing to grasp? William Masters is dead, and Virginia Johnson—communicating via her son Scott—resisted repeated wheedling, I mean requests, for an interview. (She is eighty-one, and in declining health.) But perhaps I could at least pay a visit to the penis machine.

asters and Johnson were tight-lipped on the topic of their artificial-coition machine. As far as I can tell, the most informative written material is this passage from *Human Sexual Response*:

> The artificial coital equipment was created by radiophysicists. The penises are plastic and were developed with the same optics as plate glass. Cold-light illumination allows observation and recording without distortion. The equipment can be adjusted for physical variations in size, weight, and vaginal development. The rate and depth of penile thrust is initiated and controlled completely by the responding individual. As tension elevates, rapidity and depth of thrust are increased voluntarily, paralleling subjective demand.

The only other information the pair provided had to do with the equipment also being used for testing barrier-type contraceptive devices, which made sense, and "in the creation of artificial vaginas," which made less sense. There was a footnote related to this latter bit, which made reference to two circa-1930 papers by Robert Frank and S. H. Geist, experts on the topic of vaginal agenesis. Some women, about 1 in 5,000, are born without a vagina, and some gynecologists—Frank and Geist prime among them—have made it their life's calling to give them one. Before Frank and Geist came along, this might entail fashioning an ersatz vagina out of a piece of the woman's intestine or—less (or possibly more) attractively—her rectum.

Frank and Geist thought it better and safer to simply stretch to its maximum the vaginal membrane the women had been born with. This was done by pushing in on it with a Pyrex tube★ several times a day. When "a narrow canal at least 2½ inches long" had been established, the women then widened the cavity by placing gradually larger Pyrex tubes inside their fledgling vaginas and leaving them in while they slept. Masters and Johnson realized that a few rounds with the artificial-coition machine might be a way to speed the process along.

The Frank and Geist papers, diverting though they were, did little to alter my impression of the artificial-coition machine as an artless and unarousing partner. I remained determined to see it. A 1996 A&E Television biography of Masters and Johnson mentioned the apparatus, but the producers told me they had not been given

---

★Pyrex's first visit to the human body cavity, but not its last. Because of its strength and refusal to shatter or splinter when broken, Pyrex is popular among safety-conscious dildo and butt-plug manufacturers. Ovenware is the mere tip of the Pyrex iceberg. Beakers and test tubes are often made of Pyrex, as are bongs, water pipes, and the Mount Palomar telescope mirror. Pyrex was originally invented as a lantern glass for trains; regular glass overheated and then cracked when snow hit it. It wasn't until Bessie Littleton, the wife of a Corning Glass scientist, baked a cake in a sawed-off Pyrex battery jar that the baking dish application got rolling. The Pyrex Web site includes a 1965 photograph of Bessie and husband Jesse "re-creating the kitchen discovery scene" for a Pyrex fiftieth-anniversary celebration. The company has no plans to re-create the Pyrex anal-plug discovery scene for future anniversary events.

access to it or to any subjects who'd experienced it—or, for that matter, to any penis-camera footage.

I sent another letter to Virginia Johnson's son. I told him that I only wanted to visit the artificial-coition machine or see some footage of it in action. Virginia Johnson need not even pop her head in to say good morning.

No reply. I phoned again. Scott gave me the sort of hello that wants very badly to be a good-bye. He said that any archival material—including anything relating to the artificial-coition sessions—had "probably been destroyed" to ensure the anonymity of the subjects. What about the machine? Surely it too hadn't been destroyed.

He wouldn't talk about it. He said: "We're really not interested in getting involved. *Follow?*" (I later learned from sex researcher Roy Levin that before William Masters died, he told one of Levin's research partners that the artificial-coition machine had been dismantled.)

And now you understand why I paid $20 to attend an event billed as follows:

SEX MACHINES:
Book Talk, Slide Show & Machine Play Party!

Masters and Johnson, it turns out, are not the only sex-machine game in town. An entire subculture exists, with enthusiasts all over the country trading tips on Internet listservs. With the 2006 publishing of *Sex Machines: Photographs and Interviews*, they even have their own coffee-table book. Coincidentally, a few days after the dispiriting Scott Johnson conversation, a newspaper editor I know forwarded me the press release for the *Sex Machines* event.

Now I'd be able to see, firsthand, whether and how the mechanical dick delivers.

. . .

t he Center for Sex and Culture does not court the curious passerby. No sign is posted on the outside of the building or inside the entryway. It is a nonprofit in a plain brown wrapper.* Eventually, you notice the street number, 298, on a window near the door. There is an intercom with a buzzer labeled CSC. When you ring it, a voice says simply, "Hello?" forcing you to announce that you are HERE FOR THE SEX-MACHINE EVENT. This being San Francisco—on a block where, not two minutes ago, a man in a cotton skirt descended from his flat to the sidewalk on a telescoping fire escape ladder and trotted away—no one finds the pronouncement to be worthy of especial notice.

On-time arrivals are asked to wait in the hall outside the door, because the machinists are not quite ready. Soon the line reaches all the way down the staircase. We are a mild-looking group. Based on appearances, we could be people in line at a Safeway or a Starbucks. A short man with a cane and a pencil-thin mustache has come with a conservative-looking woman in a navy and beige raincoat. One man I talk to is a physicist; one woman is a journalist. Mostly, I see couples and solo men who do not appear to be any particular type other than curious. A few gay male couples are in line, but the event, like the machines themselves, seems to lean hetero.

CSC's founder, the irrepressible and incandescent Carol Queen, a cultural sexologist, steps up to address the audience ("Ladies and gentlemen and everyone else!"),

---

*In addition to book events and art exhibits, CSC sponsors sexuality workshops and "practical skills-building events" (e.g., "Sex Club Etiquette," "Safer Sex," "G Spot Stimulation," "Positions and Toys for Plus-Sized Partners") and holds an annual fund-raiser called the Masturbate-a-Thon.

who have by now filed inside and settled into chairs. Queen introduces Timothy Archibald, the author-photographer of *Sex Machines*. Archibald is dressed in baggy orange painter pants and a red plaid shirt. He holds a bottle of Corona and occasionally reaches up to feel his shaved head. Archibald's default expression is a relaxed smile. It is easy to see why the builders, with some cajoling, agreed to pose for him. Archibald is a respected fine art photographer, and his book (really and truly) is a fine art photography book. No one was photographed using the machines; the images are portraits of the inventors, all men, in their homes and garages, posing with their machines like 4-Hers with their stock. The friend who passed along the press release also gave me a review copy of the book, and it lay around in our living room enjoying pride of place with the other art books until my husband's parents came out to visit.

The typical scenario, Archibald says, is a married guy "who likes building things." He comes across someone else's sex machine, is fascinated, decides to build one himself. "He presents it to his wife, who goes, 'Wha?' and then he sells it on eBay." Only one of the machines here tonight is manufactured for general sale.

Someone has plugged in a homemade machine on the floor behind Archibald. The motor housing is the size of a lunchbox and is raised on one end, like a slide projector. A putty-colored phallus on a stick slides quietly in and out. The erotic appeal seems limited. It would be like dating a corn dog. This is the basic setup for most of these machines and, I imagine, for Masters and Johnson's device: electric motor attached to piston attached to phallus. Or as Archibald puts it: "something out of shop class, with a human appendage stuck to it." The more sophisticated models include, as Masters and Johnson's did, a control box allowing the user to vary the speed and the depth of the thrusting. A few of the phalluses can be set to vibrate, but most just go in and out, in and out.

Archibald winds up his talk and invites questions. A woman in wire-rimmed glasses and a green T-shirt raises her hand. "What we're seeing is a lot of dildos going in and out of orifices. Given that the majority of women don't orgasm this way, do any of these machines pay attention to the clitoris?"

Archibald concedes that the machines represent a stereotypically male notion of what women enjoy. Only one machine that he knows about—it is not here tonight—attends to the clitoris. Another woman raises her hand. So what, in that case, is the appeal? "Is it the eroticization of being fucked by a machine, or the regularity of the thrusting?" Archibald looks at his builders. No one seems to have a solid answer.

William Harvey had an answer. In 1988, long before the current Internet-fueled sex-machine boom, this man obtained a patent for a Therapeutic Apparatus for Relieving Sexual Frustrations in Women Without Sex Partners. Unlike the machinists here tonight, Harvey was very clear on the purpose of his machine. "Vibrators and sex aids . . . cannot satisfy the true needs of a partnerless woman who wants not only the ultimate climax or orgasm, but also the feeling that she is actually having sex with a partner."

The partner Harvey invented took the form of a toaster-sized, unadorned metal box with a motor inside and a "continuously erect yet resiliently pliable artificial penis," a.k.a. "penial assembly," sticking out of the front of it. The box was mounted on a track, upon which it rolled to and fro, finishing each stroke with "a rapid cam-operated thrust." On some level, Harvey must have sensed that certain aspects of an actual partner were missing—warmth, say, or personality, arms and legs, a head, a soul. Harvey could not provide these things, but he could provide "the look and feel of a male's pubic hair." At the base of the penial assembly was a wide, black, wiry cuff of "fur-like or hair-like material." For the partnerless woman who wants not only the ultimate climax or orgasm, but also the feeling that she is actually having sex with a shoe buffer.

I wanted to ask Mr. Harvey how many units he has sold to partnerless women, and whether the women are sexually gratified by it. Whoever he is, he no longer lives in the town listed on the 1988 patent. A Web search turned up dozens of William Harveys. One had an email address. For the heck of it, I sent a note asking if he was the William Harvey who owns U.S. Patent 4,722,327, a Therapeutic Apparatus for Relieving Sexual Frustrations in Women Without Sex Partners. A reply arrived the following morning: "I am not the correct William Harvey, but your research does sound very interesting."

Archibald takes one last question, and then the audience is invited to mingle with the machinists. Allen Stein, inventor of an elaborate, chair-mounted nightmare called the Thrillhammer, has an answer to the appeal question. Stein, an attractive, sturdy former marine waste engineer,★ is the "chief visioneer" for a company that makes videos

---

★Company motto: "Number 1 in the Number 2 Business." And Number 45 on the list of companies that come up on Google claiming the same motto. The list includes port-a-potty rentals, septic tank emptiers, Dr. Merry's PottyPal Potty Seats, and, above all, pooper-scooper services. There are so many of these that they have their own professional group (the Association of Professional Animal Waste Specialists) with its own Statement of Philosophy. Members are obliged to "operate in such a manner as to reflect honor upon the animal waste industry" and "to place service to . . . the animal waste industry above personal gain," and you'd need a pretty big scoop for *that* load.

featuring models in flagrante with sex machines. Stein says the site appeals to men who like heterosexual pornography but are uncomfortable looking at the naked men that the naked women are having sex with. It is, he says, "porn for homophobes." Additionally, Stein says, some couples buy sex machines because the idea of a threesome appeals to them, but they are nervous about inviting a stranger into their lives. (A similar trepidation also prompts the occasional closeted gay man to experiment with a machine.) Stein puts his hand on the seat of the Thrillhammer. "Here's a third party who's safe, who's not going to take over the relationship." The Thrillhammer is eight feet tall and makes a sound like an off-balance washing machine.

"I don't know about that, Allen."

When asked whether he knew of any women who get off solely on repetitive mechanical thrusting, Stein says he doesn't keep in touch with his customers, so he doesn't really know. "Ask her," says Stein. The woman with the wire-rim glasses is preparing to board the Thrillhammer. She has changed out of the green T-shirt and into a floor-length black satin nightgown. The eyeglasses are gone. This woman, it turns out, is a friend of the CSC and something of a luminary in the San Francisco sex scene. It's possible she's participating as a favor to CSC, to try to get the thus-far-nonexistent "play party" aspect of the event rolling. Or she may simply be curious. I did not ask. Allen shows her how to work the controls, and then retreats to the sidelines to check his BlackBerry.

Very few people have noticed that a woman is poised to mount a machine, or vice versa, I'm not really sure. The audience members are standing around chatting, holding plastic cups of wine and looking at the machines as though they were sculptures at a gallery opening.

Eventually, a small throng gathers beside the

Thrillhammer. The woman in the nightgown says that she is sixty-eight years old. I would have guessed fifty. She climbs onto the machine, which is mounted on a nineteenth-century gynecologist's chair. Back then there were no stirrups, but instead long upholstered extensions where you rested your legs. She leans back and passes the control panel to a stranger, a quiet, suburban-looking woman with naturally blond hair and strappy, heeled sandals. "Surprise me," this woman is told.

Allen hands the nightgown woman what looks like a microphone. "I include one of these free with every purchase," he says to the onlookers. This might be more than I can take: a woman in her Social Security years, singing karaoke while being delved by a plunging, vibrating phallus. The woman places the microphone between her legs. It's a vibrator—a Hitachi Magic Wand.

It is more interesting to watch the woman who is manning the Thrillhammer control toggles. She is making the sort of face I make when I watch those plastic surgery shows or pull off a Band-Aid. She seems worried that she might be inflicting damage. The woman in the nightgown reports that she finds the motion of the Thrillhammer's appendage to be distracting. She has the blond woman slow it down and tries to focus on the magic of the Wand. Presently, the woman in the nightgown reaches a tidy peak. Watching her is no more erotic or awkward than watching a stranger sneeze.

To get back to the question at hand, we did not see a sex machine—at least, without help from Hitachi—escort a woman through even one complete cycle of sexual response. But there is at least one woman who has, as she put it, "had orgasms just off the machine." She appears in a photograph in Archibald's book, curled up nude on a floral couch with an unnamed machine parked on the carpeting at her feet. "I like it because I have total control," she is quoted as saying. "If it were a guy he'd be doing it for himself, his own pleasure, but this is all about me." Masters and Johnson observed this as well, noting in one of their books that "the biggest detriment for effective female response was male control of thrusting pattern." In other words, a phallus is a welcomed addition to female pleasure, as long as the woman has some say about it—its speed, its angle, its depth, its outfit.*

In Alfred Kinsey's sample of 8,000 women, 20 percent reported occasionally making some kind of "vaginal insertions" when they masturbated—though usually in addition to doing something directly clitoral. It was Kinsey's opinion that many of these women penetrated themselves because their husbands liked to watch them, or because they didn't know any better. It's more likely that these women had discovered their G-Spot (or female prostate, or front-wall erotic zone, or whatever you wish to call it), and that they weren't simply thrusting straight-on, like a penis in the missionary position. If vaginal stimulation didn't contribute in any way to women's pleasure, why would "rabbit"-style vibrators (with both a clitoral and an internal component) be the sales phenomenon that they are?

Some time later, I came across an explanation of the Masters and Johnson coition-machine mystery in a paper by feminist Leonore Tiefer, a professor of psychiatry at the

---

*In 1998, a woman in Saline, Michigan, received a patent for a Decorative Penile Wrap intended to "heighten sexual arousal of a male and female prior to intercourse." The patent includes three pages of drawings, including a penis wearing a ghost outfit, another in the robes of the Grim Reaper, and one dressed up to look like a snowman. I tried to call the examiner listed on the patent, Michael A. Brown, but he has left the U.S. Patent and Trade Office. And who can blame him.

NYU School of Medicine and a vocal critic of the medicalization of women's sexual concerns. Tiefer points out that near the end of *Human Sexual Response*, Masters and Johnson reveal that in order to be accepted as subjects, women were required to have "a positive history of . . . coital orgasmic experience." Far from being randomly selected representatives of average American womanhood, they were cherry-picked to be easily orgasmic.

Marie Bonaparte, the great-grandniece of the little guy in the wide hat, claimed to have found a simple answer to the question of why some women climax readily from intercourse alone and others don't. She found it using nothing more elaborate than a measuring tape.

3

The Princess and Her Pea

*The Woman Who Moved Her Clitoris,
and Other Ruminations on Intercourse Orgasms*

Once upon a time, there was a princess named Marie. She had long, thick curls and beautiful brown eyes, and her clitoris was three centimeters away from her vagina. This last bit was very depressing for the princess. She could never manage an orgasm during intercourse, and she felt certain that the far-off placement of her clitoris was the reason. Princess Marie—whose last name was Bonaparte and whose great-granduncle was Napoleon—was a passionate woman with a commanding libido. Yet sex left her unsatisfied. Her troubles had partly to do with her husband, Prince George of Greece, a closeted gay man,* who, she wrote in her diary, took her on their wedding night "in

---

*Marie was unaware of her prince's proclivities when they married. Her suspicions were roused by the drawings of Greek athletes that George hung on his dressing room walls and, later, by his decision to serve as the gymnastics examiner at the Panhellenic Games. Marie had just given birth to their first son and complained in her diary that while she was home all day "suckling Peter," George was off, well, suckling peter.

a short, brutal gesture, as if forcing [himself] . . . and apologized, 'I hate it as much as you do. But we must do it if we want children.'" But you could not hang the princess's discontent entirely upon the gigantic handlebar mustaches of Prince George. For intercourse with the prime minister of France also left her cold, as did intercourse with her husband's aide-de-camp and the three other lovers that she took while married to George.

Marie, who lived mainly in France, went so far as to seek scientific proof for her anatomical theory of frigidity. Bonaparte was not a physician, but she played one from time to time, and well enough to have published a paper in a 1924 issue of the medical and surgical journal *Bruxelles-Médical*. She used a pseudonym, A. E. Narjani, but readers surely sensed this wasn't the handiwork of the customary *Bruxelles-Médical* contributor. Here she is in the journal's pages, describing women like herself:

> They remain, despite all the caresses, even with all the tender gestures that should fulfill their heart, eternally unsatisfied by their bodies. Because, for these women, the moment which should bring them the greatest escape to joy, brings each time instead the torture of the ancient Tantalis. The ecstatic smile of the truly satisfied woman never shines on the face of these women tantalized by love. . . . And as happiness is for them unattainable, they are fated to pursue it, from lover to lover, in a hopeless hunt, until the knell of old age tolls.

Princess Bonaparte, working with doctors she knew, put the ruler to 243 women and interviewed them about their sex life. The subjects landed in one of three categories, based on the distance between their vagina and their clitoris. The women with the lengthiest span, a distance longer than two and a half centimeters (an inch), she labeled *téléclitoridiennes*. They made up 21 percent of her sample. Women in this category, she claimed, were incapable of *volupté*—or "normal voluptuous reaction," meaning orgasm—during intercourse. *Téléclitoridienne* means simply "female of the distant clitoris," but it had a lovely, aristocratic ring to it—calling to mind a career woman in heels and sweater set, cabling reports from her home in Biarritz. At the very least, it had a nicer ring to it than "frigid."

The luckier women were the *paraclitoridiennes* (*para*-meaning "alongside"). These women, 69 percent of Bonaparte's sample, had clitorises less than an inch distant and, she said, were almost guaranteed a voluptuous reaction from the in-and-out thrust of a penis. (This does not jibe with more modern data, which puts the figure for women who have orgasm from intercourse alone at about 20 to 30 percent.) The remaining 10 percent of Bonaparte's women, the *mesoclitoridiennes*, whose distance lay right at the one-inch cutoff point, inhabited the "threshold of frigidity." They fell upon either side, depending on their mood, their husband's compensatory skills, his feelings about Greek sprinters, and what have you.

Luckiest of all, Bonaparte wrote, are mares and cows. "Nature has favored domestic animals over womankind," she lamented in her paper, pointing out that the clitorises of these animals were "located right on the border of the genital orifice." Given that the average horse or cow liaison is over in a matter of seconds, these creatures sorely need a clitoral leg up. It's just as well Bonaparte never investigated the private bits of the domestic sow, whose clitoris sits *inside* its vagina.

. . .

If the distance is less than the width of your thumb, you are likely to come." This catchy anatomical ditty was penned not by Marie Bonaparte, but by Kim Wallen, an Emory University professor of behavioral neuroendocrinology. Wallen spends most of his time studying sex hormones at the Emory-based Yerkes National Primate Research Center on the outskirts of Atlanta, but he has of late been researching the physiology of intercourse. Wallen, who has a starring role in chapter 14, was the person who first told me about the princess and her clitoral travails. He was intrigued by Bonaparte's findings, but he did not, at first, put much stock in them, mainly because the science of statistics in 1924 was relatively primitive. Then he ran her numbers himself. The vaginal-clitoral distances, he said, turned out to perfectly predict which women would have orgasms in intercourse and which wouldn't. The cut-off point, as Bonaparte had noted, lay at around an inch—the width of a typical thumb. I asked him if he was going to trademark his "rule of thumb."

"Yes," he deadpanned. "And I'm going to start selling a little custom-made ruler."

Bonaparte includes only 43 of the 243 subjects' measurements in her paper, so Wallen's data were limited to those. Nonetheless, the data are so consistent that 43 turns out to be more than enough to say that there is, as he puts it, a very powerful, statistically reliable relationship. "Certainly strong enough to say there's something here that's worth looking at." Wallen plans to do a larger study himself, as soon as he has time.

A more recent study of genital variations among fifty women confirmed the range of distances Bonaparte found: from a half inch to almost two inches, with an average at around an inch. This study, by U.K. gynecologist Jillian Lloyd and colleagues, had nothing to do with orgasm. Lloyd sought to document the truly remarkable degree of variation in the size and shapes of women's genital features.* The hope was to reassure those who worry that their clitoris, say, is abnormally large or their pubic hair too rangy. Pornography, Lloyd points out, exposes us to idealized, highly selective images, making women needlessly self-conscious (and labia-reduction surgeons rich).

Wallen, like Masters and Johnson, thinks it's possible that a majority of the so-called vaginal orgasms being had during intercourse are in reality clitoral orgasms. But unlike Masters and Johnson, he doesn't suggest that most women are having them easily. He believes, like Bonaparte, that the women having them—the *paraclitoridiennes* of the world—are an anatomically distinct group whose sexual response is different from that of the majority of women. And that maybe these women are "where the whole notion of the vaginal orgasm originally came from."

I offered to be a statistic in Wallen's new study. He sent me detailed instructions on how to do the measuring. It's not as simple as it sounds, because Bonaparte's measurements—and thus Wallen's too—were from the clitoris to the urethra (where urine exits the body), rather than from clitoris to vagina. (The urethra is dependably close to the clitoris and makes a more precise measuring point.) A clitoris is easy to find, but urethras are sometimes hidden inside the opening to the vagina, and often hard to see. I emailed Wallen twice with questions.

"It is interesting," he wrote back, "that you could reach this stage of life and never really have any call to know how the parts line up." It is, I guess, yes. But you can't even see

---

*The width of a single inner labium, for instance, ranged from just under a third of an inch to two full inches. Gynecologist and early sex researcher Robert Latou Dickinson mentions a patient whose paired labia minora, in "fullest stretch," covered a nine-inch span.

the real estate in question without a hand mirror. I would wager that most men can better visualize a woman's vulval particulars than can most women.

If you try this yourself, I recommend doing so when no one is home. Otherwise, you will run the risk of someone walking in on you and having to witness a scene that includes a mirror, the husband's Stanley Powerlock tape measure, and the half-undressed self, squatting. No one should have to see that. It's bad enough you just had to read it. Also, put the tape measure away when you're done. My husband saw it on the bedside table and said, "What were you measuring?"

Bonaparte also discovered a correlation between a woman's height and how close together her vulval features are. Shorter women tend to have shorter spans. Wallen says the relationship is less reliable among taller women, owing to certain vagaries of puberty, but a general trend seems to exist. Then again, women were a lot shorter in Bonaparte's day; Wallen is less comfortable with this part of the equation.

By the time you are reading this, Wallen's analysis of Bonaparte's data will likely have been published. The journal editor already has plans to issue a press release. If you publish this, I said to Wallen last week, think what it will do to tall women. Are men going to hear about this and think tall women are less likely to orgasm? Wallen didn't think men would give it all that much thought. He also believes—or, at least, hopes—that the long-spanned nonorgasmic-in-intercourse woman might be relieved to learn of an anatomical (rather than psychological) explanation for her situation.

Based on a small, anecdotal survey that he'd probably prefer I not mention, Wallen has also been finding that women with small breasts seem more likely to have shorter distances. Put it all together and it spells bad news for the stereotypical American male. The stereotypical ideal female—Barbie tall with Barbie big breasts—is the one least likely to respond to vaginal penetration.

r oy Levin has a mildly different take on the intercourse orgasm debate. Levin is the founder of a sexual physiology lab at the University of Sheffield and the author of all the most eye-catching papers (e.g., "The Isolated Everted Vagina," "Wet and Dry Sex,"* "Vocalized Sounds and Human Sex"). Sadly (for me and for you), Levin has retired and so I could not watch him at work. I had to make do with watching him at lunch, which was less, but not that much less, of a spectacle. Levin's father was a butcher, and the family affinity for meat lives on. Levin lustily dispatched a calf's liver that, in my memory anyway, was as big as his shoe.

Levin ran a small investigation that focused on the erotic sensitivity of the female urethra. More specifically, his interest was in the toenail-sized patch of tissue that surrounds the outlet of the urethra: the periurethral glans. In men, this glans, which lies at the tip of the penis, is exquisitely erogenous. Levin has observed that when women have intercourse, the female glans, as he calls it, is repeatedly pulled partway inside the vagina—as much as half of it typically disappears into the crevasse with each thrust. Perhaps women who have orgasms during intercourse, he

---

*In parts of Africa, Haiti, and Indonesia, the wet, welcoming vagina is a turnoff. Men describe it as tasteless or diseased, and women insert all manner of drying agents to deliver the "dry sex" preferred by their men. "All manner" meaning: shredded newspaper, cotton, rock salt, detergent, bark, dried animal excreta. I was aghast until I read Levin's paragraph about superabsorbent tampons, which, one study claimed, can cause the outer layer of vaginal cells to dry out so severely that it peels away.

reasoned, are women who are more erotically sensitive in this spot. (Though you'd think that, here again, proximity might play a role; i.e., the closer the glans is to the vagina, the more gets pulled inside.)

I asked Levin how he had documented the disappearing glans. I pictured him in his lab with a camera and headlamp, hunched over a copulating couple. In fact, he was hunched over a TV. He took his measurements off freeze-frame images from porn video close-ups. "These guys know how to shoot this stuff," Levin said, spooning English mustard onto his plate. "They've got the good lighting, they get the angles right, they've got people who know how to direct it, you know—'Just lift your leg up a little bit, darling.'"

The next step would be to compare the sensitivity of this bit of vulval acreage in two groups of women: those who have orgasms from intercourse, and those who don't. The study would be simple enough. You'd need little more than a set of von Frey hairs.* These are boar hairs—or, nowadays, nylon threads—of differing stiffness pressed to the skin to quantify tactile sensitivity. Why hasn't anyone done this? Because almost no one gets funding for purely physiological research anymore. The grant money these days goes to studies of drugs for female sexual dysfunction (men's troubles having been more or less vanquished by Viagra and its kin). "The money's impossible, unless you've got an obvious application," said Levin. "They want to know, How's it going to help patients?"

In this case, of course, it wouldn't help them much. If the sensitivity of the female glans turns out to make the

---

*Max von Frey was an Austrian physiologist who invented a heart-lung machine some time before an American named John Gibbon did. Dr. Gibbon ultimately became known as the machine's inventor, while von Frey's name was linked to calibrated pig hairs. Life is unkind.

difference in who has orgasms from intercourse, then, as Levin says, "anatomy may well be destiny."

m arie Bonaparte was not willing to accept her destiny. Drastic measures were undertaken. *The princess had her clitoris moved.* The relocation was carried out by a Viennese surgeon of Bonaparte's acquaintance, Josef Halban. It was apparently her idea. A later paper by her shows photographs of the procedure—blessedly obscured by the poor quality of the photocopy I have—and she has labeled it the Halban-Narjani operation. The "simple" procedure, as Bonaparte called it in her paper, entailed cutting the organ's suspensory ligaments,★ freeing it to be stitched in place slightly lower. She must have believed it was simple, because she allowed Halban—after an initial test run on a cadaver—to perfect his technique on her. Given that much of the clitoris is hidden below the body's surface, moving it was perhaps not as simple a matter as Halban had thought.†

★A trick also employed to lengthen penises. Though an organ thus freed tends, when erect, to veer off sideways and/or hang down dispiritedly. Anyone considering this should know that an Italian study done in 2002 found that absolutely everyone in a group of 67 men seeking surgical penis lengthening had penises within the normal range (from 1.6 to 4.7 inches, flaccid). "Normal" was determined, rather dauntingly, by measuring 3,300 Italian military recruits. The authors blame abnormally endowed porn stars for the growing epidemic of penile insecurity.

†Far easier to move a neoclitoris, which is the technical term for a transgender woman's clitoris, typically fashioned from a stitched-in-place nub of penile glans tissue. Do transgender women ever request a closer placement? No, says Harold Reed of the Reed Centre for Genital Surgery in Miami. "They are going for looks." As in, trying to look natural. To that end, Reed places them one inch above the urethra, right smack at average.

Though Bonaparte writes that the operation was a success in two later patients, it did nothing for her. Some years later, Halban offered to redo the procedure. Bonaparte took him up on it, and again was disappointed.

All the sadder, given that Bonaparte, in her own paper, outlines a perfectly workable nonsurgical solution to the *téléclitoridienne*'s predicament: Try a different position. "Only a change in position during the embrace—the best being the face-to-face seated position—which forces contact between the clitoris and the male organ can give the *téléclitoridienne* the experience of simultaneous pleasure that other women enjoy." Sadly, Prime Minister Briand was a man with little discernible lap.

Had Bonaparte paged through any one of the "marriage manuals" popular at the time, she would have found still more promising coital configurations. Sexually, the 1920s and '30s were an oasis of openness and common sense between the twin deserts of Victorian repression and fifties-era conservatism. Jessamyn Neuhaus, in an article about marriage manuals in the *Journal of the History of Sexuality,* credits the birth of the birth-control movement for this change. Having spent the past century shackled to reproduction, sex suddenly emerged as a recreational pastime. Orgasm—particularly women's—became a prerequisite for good health and matrimonial harmony, and dozens of authors, medical and not, chimed in with tips on how to achieve it.

The best known of these authors was Dutch gynecologist Theodoor Van de Velde. *Ideal Marriage: Its Physiology and Technique* was the first in his trilogy of advice manuals aimed at nurturing "the perfect flower of ideal marriage" and "combating the forces of mutual repulsion"—the latter including fermented clitoral smegma ("extremely disgusting"), bad breath,* and purulent rhinitis. The book, some of whose European editions went through more than forty printings, includes twenty-five pages on coital positions and directly addresses the téléclitoris: "When the clitoris is set very high the woman must take an attitude which accentuates pelvic inclination." Yes, but inclined in which direction? Van de Velde went into admirable detail about the different positions and their pros and cons vis-à-vis orgasm and fertility, but his Twisteresque descriptions cry out for illustration. The Anterior-Lateral Attitude, for example, has the woman "half-lateral, half-supine, with a corresponding half-lateral, half-superposed attitude of the man, which is possible by appropriate arrangement of pillows." Just the names alone are daunting. Faced with "Second Extension Attitude: Suspensory (Variation [b])," all but the most motivated couple might well throw in the towel and let the damn perfect flower wilt. Alas, as relaxed as the era was, it was not ready for mainstream books with pictures of people having sex. Even with no illustrations,

---

*Which does not necessarily include semen breath. Van de Velde claims that a "slight seminal odor" can be detected on a woman's breath within an hour after intercourse, and that the effect can be very arousing for the man. Or anyway, the man who enjoys smelling semen. Van de Velde's semen connoisseurship must surely have raised some eyebrows: "The semen of the healthy youths of Western European races has a fresh, exhilarating smell; in the mature man it is more penetrating. In type and degree this very characteristic seminal odour is remarkably like that of the flowers of the Spanish chestnut, which . . . are sometimes quite freshly floral, and then, again, extremely pungent. . . ." Perhaps Van de Velde's bitter outlook on the typical marriage ("that morass of disillusion and depression") was traceable to something deeper than his distaste for his first wife.

the Catholic Church put *Ideal Marriage* on its list of prohibited books.

The first to prevail on this front was our hero, the gynecologist-turned-sex-researcher Robert Latou Dickinson. His *Atlas of Human Sex Anatomy* includes a two-page spread of fourteen thumbnail Coital Diagrams with terse, pronoun-sparse titles: "Pillow Lifts Hips," "He Diagonally Across." Chaste as these drawings are—the bed beneath the couple is drawn in more detail than are their faceless, featureless bodies—they were edited out of the first edition. Dickinson tried to appease his publisher by replacing the human forms with a pair of entwining robots; however, he reports, "these evasions proved to be not a little absurd" and the publisher eventually relented.

Not content to sketch the hows of what he called vulvar orgasm—meaning an orgasm that arrives during intercourse but owes its existence to clitoral involvement—Dickinson dug into the whys. Figure 147 shows disembodied penises rubbing against little button-mushroom clitorises and stretching them to and fro. The key, explained Dickinson, was to find a position in which the man's pubic bone pressed against the woman's clitoris and/or moved it to and fro. The end points of the to-and-fro, which he termed "excursion," are illustrated in such a manner that the vulva, at first glance, appears to have three clitorises, as though perhaps Josef Halban had been mucking about.

Dickinson, carlengths ahead of his time, was a champion of "Woman Above" postures, and included three in his Coital Position pages. "Woman, if above, can regulate excursion and pressure," says the Dickinson caption, striving to be as untitillating as possible.

Unlike Marie Bonaparte, marriage manual authors of the era saw no reason for a woman—*téléclitoridienne* or other—to limit the tools of sexual gratification to her husband's penis. Van de Velde was an advocate of cunnilingus—as long as it was a prelude to, and not a substitute for, intercourse, and as long as the "spotless cleanliness and wholesomeness of the bodies" could be counted upon. Even then, he sort of gets to it by the side door, in a passage on vaginal lubrication:

> The most simple and obvious substitute for the inadequate lubricant is the natural moisture of the salivary glands. It is always available; of course it has the disadvantage of very rapid evaporation. . . . During a very protracted local or genital manipulation, this form of substitute must be applied to the vulva, not once, but repeatedly. And this may best, most appropriately, and most expeditiously be done without the intermediary offices of the fingers, but through what I prefer to term the *kiss of genital stimulation,* or *genital kiss:* by gentle and soothing caresses with lips and tongue. . . . The acuteness of the pleasure it excites and the variety of tactile sensation it provides, will ensure that the previous deficiency is made good.

In his advice for the high clitoris owner, Van de Velde references the "obvious method of combining vaginal friction by the phallus with simultaneous clitoral friction by the finger." Neuhaus quotes the 1935 edition of *Sex Practice in Marriage*: "Should a man be unable to restrain himself and have an orgasm before his wife, he *must* keep up the clitoris stimulation until his wife has reached the climax." It was a good time to be a woman. They got the vote, they got birth control, and now they had husbands who gave genital kisses and finger friction. The stage was set for Alfred Kinsey. The survey results presented in his 1953 book

Sexual Behavior in the Human Female gave vaginocentrists a drubbing from which you'd imagine they would not easily recover. Only a third of his subjects reported easily and consistently having orgasms from intercourse, and those that did were benefitting from clitoral stimulation by the man's organ or body. Ninety-five percent of them said that their husbands practiced "manual stimulation" before the couple got down to the business of intercourse.

But times had changed. The fifties were not the twenties. *Sex Behavior in the Human Female* was met with a hail of outrage and criticism. American manhood would not abide the sexually sophisticated (i.e., demanding) woman, and it fought back hard. Among the more vocal vaginal crusaders was Arnold Kegel, inventor of the eponymous pelvic squeezing exercises. (Dr. Kegel originally prescribed the exercises as a remedy for incontinence, but patients began reporting a happy side effect: They were having orgasms during intercourse where none were had before.) "I believe, as many others do, that research in the physiology of the [pelvic floor] muscle[s] has produced overwhelming evidence in favor of the existence and importance of vaginal orgasm," harrumphed Kegel in a critique of Kinsey's book in a letter to the editor of the *Journal of the American Medical Association* (*JAMA*). Kegel added that the one-third of women who reported to Kinsey that they could not climax during intercourse were simply afflicted with flabby pelvic-floor muscles. They need only take up Kegeling—up to sixty squeezes, three times a day—to enjoy, if they had any spare time left, the vaginocentric ecstasies of their better-toned sisters. "Concentration on the muscle seems to cause patients in this group to forget the clitoris," said Kegel, in a direct inversion of Kinsey, who wrote that once his interviewers had explained the effectiveness of clitoral stimulation to women who had confessed to masturbation by "vaginal insertion," they dropped the latter.

It seems Kegel didn't read Kinsey's book all that closely. Kinsey may have dethroned the vagina, but he didn't kick it out of the castle. Item 5 on his list of "six or more sources of the satisfactions obtainable from deep vaginal penetrations" is stimulation of the pelvic-floor nerves. Though it is true that Kinsey gives more weight to item 3—"stimulation by the male genitalia or body pressing against . . . the clitoris" and the rest of the vulva.

The backlash to Kinsey and the general tide of conservatism turned the passive, vaginal orgasm into the holy grail of female sexuality, "the hallmark," wrote Carolyn Herbst Lewis in the *Journal of Women's History,* "of a well-adjusted and normal femininity." The attitude was evident in the new crop of marriage manuals, now circumcised of their clitoral references, and even in the medical journals of the day. Lewis writes that some doctors used the newly sanctioned premarital medical examination—a venereal-disease-prevention measure enacted in a majority of U.S. states during the fifties and sixties—to prepare women for a life of completely fulfilling intercourse. A woman was "assured that her responses could be as full and satisfactory as her husband's" yet given no advice about clitoral foreplay. The chief obstacle to the vaginal orgasm, in the physicians' minds, seemed to be penetration anxiety. Journal articles counseled doctors to administer, in cases of thick and ornery hymens, some pre-wedding-night snipping or stretching during the premarital exam. The latter could be done with their fingers or, as one 1954 *JAMA* author counseled, a "well-lubricated Pyrex centrifuge tube." Again with the Pyrex tubes.

Marriage manuals and premarital exams had long disappeared by the time the front wall of the vagina and its "G-spot" gained widespread status as an erogenous zone. Otherwise, you might have found a curiously unconservative tidbit of advice for women bent on having intercourse orgasms: Try it doggie style. Zwi Hoch, of the Center for Sexual Therapy at the Rambam★ Medical Center in Haifa, Israel, published a paper in which he trained 64 percent of a group of 56 noncoitally orgasmic women to have orgasms by stimulating the front wall of their vagina. While most were using their finger, some had managed it with "anteriorly directed intercourse." Ernst Grafenberg†—the gynecologist who, in 1950, first wrote of "an erotic zone" on the front wall of the vagina "along the course of the urethra"—also advocated sex "*a la vache*" (like the cow) as a better way to hit the spot. "The stimulating effect of this kind of intercourse," he wrote, "must not be explained away . . . by the melodious movements of the testicles like a knocker on the clitoris."

The premarital doctor's visit was not an entirely bad idea. There are some couples for whom a quick inspection and some frank chat was clearly in order. For example, Robert Latou Dickinson writes that he encountered, over his many decades of clinical practice, eighteen women whose virginity had remained intact despite having (what they mistook to be) intercourse for years. "The husbands and wives, though otherwise intelligent, thought the cleft of the vulva was as deep as his organ was expected to go." Then there was the woman written up in a 1965 issue of *JAMA* whose husband was mistaking her urethra for her

---

★No one in Israel titters over the seeming irony of a sex therapy center in a hospital called Rambam. Rambam is short for the Rabbi Moshe ben Maimon (a.k.a Maimonides). Though I now associate him with rear-entry intercourse, the Rambam, as he is known there, was an important medieval Jewish philosopher.

†The "G" in G-spot stands for Grafenberg. The term was coined by G-spot popularizer and researcher Beverly Whipple. Her colleague's suggestion to name it the "Whipple Tickle" was ignored, making life easier for her children and other innocent Whipples.

vagina. By the time a doctor discovered what was going on—she was, understandably, having some continence issues—her urethra had been stretched so far that it "readily admitted two fingers." The author of the paper found thirteen such cases reported in various medical journals, all attributable to rigid, unyielding hymens.

You could not blame a tough hymen in the case of the gay man who—he told Kinsey collaborator Wardell Pomeroy—had intentionally stretched his urethra to accommodate his lover's penis.

even Marie Bonaparte succumbed to vaginal propaganda. In Bonaparte's case, the change of viewpoint coincided with a midlife career move. In 1925, a year after publishing the *Bruxelles-Médical* paper, she met Sigmund Freud and decided to become a psychoanalyst. Freud was no friend of the clitoris. Freudian theory holds that grown women who derive their sexual satisfaction from their clitoris are stuck in a childlike state. This "phallic" phase is supposed to end at puberty, when a woman embraces her proper role as a passive, feminine being. "With the change to femininity," he wrote in *New Introductory Lectures in Psychoanalysis*, "the clitoris should wholly or in part hand over its sensitivity, and at the same time its importance, to the vagina."

Bonaparte had to do some backpedaling on her clitoral placement theory. For if, as Freud insisted, the well-behaved clitoris relinquishes its sensitivity to the vagina at puberty, then in theory a woman shouldn't need a clitoris for sexual gratification, let alone care where it's placed. In her 1953 book *Female Sexuality*, she discredits her *Bruxelles-Médical* article, saying she had since encountered both frigid *paraclitoridiennes* and *téléclitoridiennes* with no problems achieving *volupté.* (Robert Latou Dickinson, in his *Atlas of Human Sex Anatomy,* also mentions having encountered exceptions to Bonaparte's "sweeping statements.")

Obviously torn, Bonaparte tried to gain a better understanding of the matter by interviewing women whose clitorises were beyond distant—they were removed, or anyway, the protruding bits were. "Are African women more frequently, and better, 'vaginalized' than their European sisters?" she wondered in *Female Sexuality.* Freud got her interested in this. He had told her that cultures that remove clitorises do so in order to further feminize the woman.* (The more common belief is that it is done to quash sexual pleasure and desire and keep women from committing adultery.)

In 1941, during the German occupation of France, Bonaparte and her family were evacuated to Egypt, and she had her chance to talk to a couple of women who had had clitoridectomies. The women were not, in fact, fully "vaginalized." Both women—though they did report having orgasms from intercourse—still masturbated clitorally, on their scars. Probably because (more on this to follow) the majority of the organ is hidden below the deck.

Purely as an aside, Bonaparte needn't have gone to Africa to find women to talk to. American women were given clitoridectomies from the 1860s up until the turn of the century. The practice was started in London, in 1858,

---

*As an alternative, one culture cuts the nipples off boys, to masculinize them. Bonaparte gives a quote obtained from a nipple-less Janjero tribesman by an anthropologist named Cerulli. "We do this because we do not wish to resemble women in any way." A journal search turned up no mention of this practice; however, the Janjero were described as fierce hunters rumored to dally in human sacrifice, so presumably nipple-hacking would have been mere fluff to them.

by a well-respected obstetrician-gynecologist named Isaac Baker Brown. Brown put out a book stating that masturbation—in women, that is—caused hysteria, epilepsy, and "idiocy." Excising the clitoris, he stated, was the only sure cure. Often he wouldn't tell patients exactly what he was planning to do to them. When Brown's colleagues got wind of what he was up to, they voted to expel him from the Obstetrical Society of London, and his reputation swiftly disintegrated. Happily, most of his patients went right on masturbating the way they always had.

In the meantime, alas, the practice had spread to the States, where gynecologists had of late gone scalpel-happy, working out one new surgical procedure after another on indigent women—without telling them they were guinea pigs. (A long-standing tradition, says historian Ben Barker-Benfield. The much-revered obstetrician-gynecologist Marion Sims, Barker-Benfield writes, purchased slaves with vaginal fistulas* as surgical practice material for a fistula procedure he wanted to try. One poor woman was given thirty unwarranted gynecological operations.)

In the end, Marie Bonaparte concluded that some women were simply born with a clitoral orientation and some with a vaginal, and there wasn't a thing to be done about it. Neither surgery nor psychoanalysis, it turned out, could fix what ailed Marie.

· · ·

---

*A fistula is an unwanted passageway that develops between two normally separate body cavities. Like tenors, there are three well-known vaginal fistulas. The Pavarotti of vaginal fistulas is the vesicovagino, linking bladder to vagina and allowing urine to dribble out where it oughtn't. Ditto the urethral one. Most odiously, there is the rectovaginal fistula (an occasional complication of childbirth), which allows flatus and feces to leak out of the vagina. Nothing to sing about.

lfred Kinsey had the most sensible take on the intercourse orgasm conundrum. Sure, it may make a difference how your clitoris is situated. And, yes, some positions are more promising than others. But what matters more, Kinsey concluded, is one's level of engagement in the proceedings. Kinsey believed the erotic responsiveness of a woman on top was not a mere matter of "anatomical relations." He made the point that "the female who will assume such a position is already less inhibited in her sexual activity." And it is the person on top who's in control—making the movements and controlling their speed and depth and direction. "In the younger generations," Kinsey wrote, "there is an increasing proportion of the females who have become aware of the fact that active participation in coitus may contribute not only to the satisfaction which the husbands receive, but to their own satisfaction in coital activity." Maybe Marie Bonaparte just never got into it.

Kim Wallen, who recently began interviewing women about intercourse and orgasms for a new study, has been finding what Kinsey said to ring true. "Women who routinely have orgasm in intercourse without explicit clitoral stimulation all say that it makes little difference what the guy does, as long as he doesn't come too soon," Wallen said in an email. Meaning, it's the women's own movement that matters most. "In fact it is sometimes preferred that he just lie there and anchor the woman's pelvis to his. The movie image of wild abandoned thrusting seems to have exactly the opposite of the intended effect in these women." Well, yes and no, I replied. Sometimes they want that. Later, you know, toward the end. There were days, talking to Wallen, when I sensed he was nearing the end of his orgasm-in-intercourse (or O-in-I, as he'd shorthanded it) rope. "The only conclusion I feel sure of at this point," he mused, "is that women are too complicated."

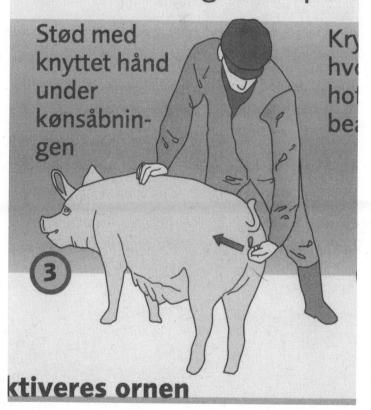

Stød med
knyttet hånd
under
kønsåbnin-
gen

Kry
hvc
ho
bea

③

ktiveres ornen

4

The Upsuck Chronicles

*Does Orgasm Boost Fertility,
and What Do Pigs Know about It?*

he inseminators wear white. Their coveralls are white
and their boots are white, and they themselves are white
too, it being the tail end of a long, dark winter in Denmark.
Their names are Martin, Morten, and Thomas, and they
have twenty sows to inseminate before noon. An informal
competition exists among the inseminators of Øeslevgaard
Farm, I am told—not to inseminate more sows than anyone else, but to inseminate them *better*. To produce the
most piglets.

To win requires patience and finesse in an area few
men know anything about: the titillation of the female pig.
Research by the Department for Nutrition and Reproduction at Denmark's National Committee for Pig Production showed that sexually stimulating a sow while you
artificially inseminate her leads to a 6 percent improvement in fertility. This in turn led to a government-backed

Five-Point Stimulation Plan for pig farmers, complete with instructional DVD and color posters to tack on barn walls. It also led to a certain amount of awkwardness on the part of Danish pig farmers, most of whom do inseminations themselves. (Gone are the days of the roving Boar Truck, a man and a pig who drove the length and breadth of Denmark, servicing sows—and unwittingly spreading pig disease from farm to farm.)

Martin, Morten, and Thomas are in the break room, eating bread with jam and drinking coffee from a slim steel thermos. They are uncomfortable speaking English, and I speak no Danish. We are dependent on Anne Marie Hedeboe, a visiting pig production researcher whose colleague Mads Thor Madsen drafted the Five-Point Stimulation Plan for sows. The mood in the room is a little starched. I called Morten Martin. I referred to the owner of the farm as "Boss Man," which sounds like the Danish for "snot." Unspoken questions hover in the air: *Do you find it arousing to stimulate a sow? How often are young male farmworkers caught getting fresh with the stock?*★ For their part, the inseminators must be wondering why on earth I've come here.

I could not adequately explain to them, but I will explain to you. Please don't worry. This chapter is not about pig sex. It is about female orgasm and whether it serves a purpose

---

★I'm too polite to ask, but Alfred Kinsey wasn't. In Kinsey's landmark 1940s survey of American men, 26 to 28 percent of college-age rural males copped to having had "some animal experience to the point of orgasm." In a few farm communities "where social restraints on this matter are less stringent," the figure jumps to 65 percent. Calves, burros, and sheep are the preferred partners, probably because they're the right height. There is no specific mention of pigs, except to say that "practically every other mammal that has ever been kept on the farm enters into the record, and a few of the larger birds, like chickens, ducks, and geese."

outside the realm of pleasure. What is accepted dogma in the pig community—that the uterine contractions caused by stimulation and/or orgasm draw in the sperm and boost the odds of conception—was for hundreds of years the subject of lively debate in medical circles. You don't hear much these days about uterine "upsuck"—or "insuck," as it was also known—and I'm wondering: Do the pigs know something we don't know?

t he job of a production pig is to produce more pigs, as many pigs as is pigly possible. The sows of Øeslevgaard shuttle back and forth between the "service" (insemination) barn and the open-floored nursing and weaning barns, where they sprawl flank-to-flank, a mounded porcine land mass. Anne Marie and I are standing around in the insemination barn. Here the sows are briefly confined in narrow pens separated by metal railings. It's like living inside a shopping cart. They seem to be in good spirits nonetheless. This may have to do with boar No. 433, a brown and white Duroc with testicles as big as a punching bag.

Thomas has hold of No. 433's tail, steering him from behind into a large enclosure that flanks the pens of the twenty sows in heat. No. 433 is a "teaser boar." His presence in the barn primes the sows for what's to come. It is not a quiet presence. The grunts of a sexually aroused boar are a soundtrack from a horror film: the deep, guttural, satanic noises of human speech slowed way down on tape. When I replayed my cassette, weeks later, I tried speeding it up to see if it would sound like speech. Perhaps I would decipher the secret language of pigs. It just sounded like someone retching.

The boar moves along the row of sow snouts protruding through the bars, rubbing each one with his own. "This is what he does," yells Anne Marie over the grunting and the banging of metal grates. "He slobbers on them." Boar saliva has a pheromone—a chemical that primes a sow in heat for mating. Strictly speaking, you do not need a boar, because you can buy a Scippy*—a remote-controlled plastic boar doused with Boarmate synthetic boar odor spray. Anne Marie's coworker Mads, who resembles a Danish Javier Bardem, if that is possible, told me about them. Mads has an endearing affection for absurdity, which must serve him well in this line of work. He dug up a picture of Scippy on his computer while I was there. "See? He's nice and pink, and he goes on wheels. It's very nice. He just has to smell and grunt. He has an MP3 player." Then Mads sank a little lower in his chair. "We bought one and tried him, and he didn't work, and the farmers didn't want him." Scippy lives in the closet now.

Number 433 has a politician's deft timing, staying just long enough to make each contact seem personal, but never lingering so long that the other sows lose patience. They seem not to mind that a viscous, salival froth clings to the boar's chin like a Santa Claus beard.

The smell of the boar is heavy and repugnant. Anne Marie said that when she flies back from farms in northern Denmark, businessmen coming down the aisle often ask if they may sit beside her. Anne Marie is young and pretty. "I say, 'It's alright, but I smell.' They think it's a joke, and they sit down." And regret it the whole way home.

Anne Marie has short, mahogany-hued hair set off

---

*Or, if you lack a sense of fun, just a can of Boarmate. One two-second spritz at the snout does the trick if the gal's in heat. Until very recently, you could go on the Web and download a Boarmate Audio File, where you could hear a British man talking very seriously, in an Alistair Cooke sort of way, about "boar odor spray."

by a pair of stylish green eyeglasses. We too are wearing coveralls and boots, to protect the pigs from any pathogens on our clothes and to protect our clothes from the smell. A full day after leaving the farm, I realize that a noticeable pall of boar stink clings to my pen and notepad. Anne Marie must wash the enviable glasses after each visit. When we go to lunch, a couple seated behind us gets up and moves.

Anne Marie's beauty and style belie a down-and-dirty education in the particulars of practical AI (artificial insemination). She has milked a boar of his prodigious ejaculate—over two hundred milliliters (a cup), as compared to a man's three milliliters—and she has done it with her hand. For, unlike stallions and bulls,* boars don't cotton to artificial vaginas. (In part, because their penis, like their tail, is corkscrewed.) AI techs must squeeze the organ in their hand—hard and without letup—for the entire duration of the ejaculation: from five to *fifteen* minutes. "You should see the size of their hands," she says, of the men and women who regularly ejaculate boars.

Anne Marie's training also covered the use of artificial vaginas (for bulls). As the bull mounts a dummy cow—a sort of heavy-duty ironing board with hair—a technician, seated alongside, quickly slips the organ into a handheld artificial vagina. It is important to stay focused. "Our

---

*Upon whom one may, if money is no object, use a Nasco Master Artificial Vagina, the "finest artificial vagina ever offered," featuring "just right stiffness." Or, for a stallion, the Nasco Missouri-Style Equine Artificial Vagina. ("The leather case permits the vagina to be carried easily. . . .")

Men, of course, are simply handed a magazine and a cup. There is a synthetic human vagina—called a syngina—but it is used in tampon R&D. Adman Jerry Della Femina, who once worked on a tampon account, joked in his book that if your campaign went especially well, you got to take the syngina to dinner.

instructor was talking to us, not really paying attention, and the bull mounted his coat sleeve." Fortunately for the man and his dry cleaner, the bull ejaculates a scant eight milliliters.

Martin, Morten, and Thomas are making sure the twenty sows are still in heat. A simple indicator is the ears, and whether they are standing up straight. Normally, they flop forward and there is a coy sweetness to the way a sow peeks out from under. (Apologies to the margarine people, but the Blue Bonnet lady comes to mind.) The other way to tell is to sit on her back; if she lets you, she's in heat.

The steps of the stimulation plan mimic the boar's rather uncouth courtship behaviors. Martin places his hand—though the boar would use his snout—at a sow's inguinal fold, the place where the back leg meets the belly, and then lifts her an inch or so off the ground. He does this four times. Boars are not gentle and so the inseminators are not either. Martin hefts the sow and lets her drop, bouncing her up and down as though testing her shocks.

Now he moves around behind her. He pushes rhythmically just below the pink fleshy bulge that is her vulva. Again, the boar would be using his snout. Martin and Thomas use their fist. Morten, who is working up the sow two stalls down, is using his knee.

"Morten?" I want to ask him if he feels odd doing this.

Anne Marie leans over to speak in my ear. "That is Martin."

Martin indicates that maybe he did at first, but he does not now. As Kaj, the farm's owner, said earlier, "It's just how it is." There is, however, a limit to what a pig farmer will do in the name of higher farrowing rates. I asked Anne Marie whether they had tried stimulating the sow's clitoris.

"We thought of trying this. But actually it was a big hurdle just to get farmers to touch underneath the vulva."

And in pigs, lucky pigs, the clitoris is *inside* the vagina. "So we thought, let's not mention the clitoris right now." However, it is possible to purchase a specialty sow vibrator. The Reflexator is made by a Belgian agricultural supply company called Schippers. Mads keeps one hidden behind a row of binders on his bookshelf but brings it out with almost no prodding. The design was recently modified so that no insertion is required of the farmers, who tend, even in a progressive country like Denmark, to be conservative. A hook now allows the Reflexator to hang benignly on the sperm feeder tube; the vibrations travel along the tube, so that hopefully, quoting Mads, "nobody has to feel funny." Except that they do. Mads estimates that fewer than 1 percent of Danish pig farmers Reflexate their sows.

The original stimulation plan had six steps, not five, but the last has been deemed, says Anne Marie, "too much." The training video—but not the poster—includes a shot of a handsome, suntanned Dane lying on a sow, his chest pressed to her back. With one hand, he reaches down beneath her to rub her mammaries and squeeze her teats.★ A close-up highlights a gold wedding ring, as though to reassure the viewer that nothing untoward is going to happen between these two.

t he linking of sexual delight and fertility, for right or for wrong, dates as far back as Western medicine itself. Hippocrates, the famous Greek "Father of Medicine," believed that female orgasm was linked, like men's, to a bursting-forth of seed—in this case, deep within the unfathomable

---

★One of the less prominently known similarities between pigs and men: They both fondle breasts. No other males on the planet regularly do this.

female interior. The commingling of male and female seed was thought to spark conception. No orgasm, no babies. Then along came that other famous Greek, Aristotle, to make the point that it is altogether possible for women to get pregnant from an interlude that did not culminate in orgasm. You never see anything written about Mrs. Hippocrates or Mrs. Aristotle, but I'd put a few drachmas on the former being the one with the spring in her step.

Happily for Western women, it was the Hippocratic version that stuck. Even long after the concept of womanly seed had been debunked, the notion that simultaneous orgasm bettered the chances of conception lingered on. It made a great deal of intuitive sense. If the man's climax was essential to the makings of new life, surely the woman's was similarly invested. Indeed, for centuries, physicians routinely advised men on the importance of pleasuring their wives. Marriage manual author Theodoor Van de Velde quotes an imperial physician's advice to eighteenth-century Habsburg empress Maria Theresa, who was slow to conceive: "I am of the opinion that the vulva of Your Most Sacred Majesty should be titillated for some length of time before intercourse." Evidently, it was sound advice; she eventually had sixteen children.

Ironically, given the goings-on in the swine farm today, it was artificial insemination that put the formal kibosh on medically sanctioned titillation. In 1777, inquisitive Italian scientist Lazzaro Spallanzani "chose a bitch of moderate size, . . . confined her in an apartment and kept the key" so as to prevent insemination of the more usual variety. Twenty-three days later, when she was clearly in heat, he "attempted to fecundate her artificially in the following manner. A young dog of the same breed furnished me, by a spontaneous emission, with 19 grains of seed." The semen was syringed into the vagina of the bitch, who, sixty-two days later, "brought forth 3 lively whelps."* The experiment raises important questions. Who agreed to let Spallanzani lock a stray dog in their apartment for twenty-three days? Did he really expect us to buy the bit about the male dog spontaneously ejaculating? But the main point here is that the needle of a small syringe is unlikely to cause a bitch much pleasure, and thus pleasure could be assumed irrelevant—or certainly not necessary—when it came to conception.

In 1840, a different dog, breed unknown, reopened the case file on orgasm as an aid to conception. A German anatomist named Hausmann killed a bitch while she was mating and then—presumably allowing a moment or two to disengage the flummoxed male from the proceedings—picked up his scalpel and opened her up. Though the male had ejaculated only moments before, semen had already reached the uterus. This suggested that something other than the sperms' lashing tails was propelling them through the reproductive tract. Uterine contractions seemed a likely candidate. Since these contractions are a hallmark of orgasm, Hausmann supposed that it was some version of canine bliss that served to suck the sperm through the cervix and into the uterus. It makes intuitive sense: Orgasm causes a release of oxytocin—the "joy hormone," also

---

*I love eighteenth-century science writing, because the humanity of it—the exhilaration of discovery and triumph—had not yet been stripped away. Check out Spallanzani's next line: "Thus did I succeed in fecundating this quadruped; and I can truly say that I have never received greater pleasure upon any occasion since I have cultivated experimental philosophy." You can practically hear the champagne corks popping, the whelps yipping underfoot.

involved in nursing—and oxytocin is known to cause uterine contractions.

Five years later, a second dog experiment confirmed what Hausmann had found, as did an 1853 guinea pig experiment, a 1930 rat experiment,* a 1931 rabbit experiment, and a 1960 golden hamster experiment.† In all cases the sperm were found to have traveled, within seconds or at most a few minutes, a distance that would take the sperm considerably longer to traverse under their own steam. But since none of these animals can be assumed to possess the same anatomical responses as a woman, the studies are inconclusive as regards human fertility.

The other trouble with these studies was that none of these researchers could say for sure that it was the female's sexual bliss that was causing the contractions that moved the sperm along, rather than some other facet of the mating process. Human semen, for instance, contains a hormone called prostaglandin, which causes powerful contractions when it comes in contact with a woman's uterus. (For this

---

*Since a male rat will mount and dismount many times before he ejaculates, this experiment required intimate knowledge of rodent sexuality. An unnamed junior staffer trained himself to "infallibly" recognize what he called the Sign of Ejaculation. Bearing no relation to the Sign of the Cross and only a passing similarity to the Mark of Zorro, the Sign of Ejaculation is a "deep final thrust." This is followed by "a period of inertia" during which the male lies collapsed on the back of the female, provided, that is, that she hasn't been plucked out from under him in the name of scientific inquiry.

†I find it hard not to project a sliver of sadism upon the scientists. The hamster guys are especially easy to mistrust, having stated in their paper that "the mated female was killed by a blow on the back of the head." *Who clubs a hamster?* What would you even use to deliver "a blow" to a head that small?

reason, when fertility doctors place sperm in a woman's uterus, they use à la carte sperm, "washed" of its semen.)

This is why a team of fearless Illinois researchers, in 1939, took to sexually stimulating rabbits themselves; they wanted to take semen out of the equation. The stimulation was done, they wrote, "with the finger," a phrasing that seems to suggest a piece of specialty lab equipment or a disembodied digit of unknown ownership rather than the flesh and blood of the researcher himself. Happily, these rabbits' lives were spared, for the observations were made via fluoroscope. Before the stimulating commenced, a dye was injected into the rabbit's vagina. This was then seen on the fluoroscope screen, poststimulation, to have spread upward into the uterus within two to five minutes.

The best animal evidence that sexual responses produce uterine contractions comes from the old water-balloon-in-the-cow study. In 1952, a different team of Illinois researchers inserted the thumbs of latex gloves into the uteri of four cows and filled them with water. The balloons were linked to a device that registered the movements of the cows' uteri as pressure on the balloon. As the researchers expected, powerful uterine contractions were detected after a bull was brought in to "mount, copulate, and ejaculate" (a process clocked, tragically, at under five seconds). More surprising was that, with all four cows, the machine began registering uterine contractions the moment the bull walked into sight.

What does this mean? Does a cow have a mild orgasm when she merely casts her gaze upon a bull? Can we be sure these uterine contractions imply orgasm? Does anyone know for sure that female animals have orgasms?

Let us check in with the sows.

. . .

t he inseminators are working side by side. Morten is lagging behind slightly. He is just now inserting a semen feeder tube into the pig he's working on. He wears the look of concentration and vague worry that my husband wears while snaking the bathroom drain. He tugs gently, making sure the placement is good and then holds up the semen pouch like an IV bag. Next he climbs on the sow's back. This is intended as a substitute for the weight of the boar. He bounces slightly to mimic the male's movements. All three men are now sitting on the backs of a sow. They look like people on an antique merry-go-round where everyone, not just the last two to board, has to ride a pig.

One by one, the semen bags are drained. It happens abruptly, sometimes after a few seconds, sometimes after a few minutes. Patience is key. You never rush a sow. (For this reason, there is no clock in the insemination area.) Thomas is said to have a way with the sows. His focus is unwavering. He never talks on his cell phone, as Kaj sometimes does. He pushes the sow's mammaries with his boots and rubs the sow's neck and behind her ears, though these things are not prescribed on the poster.

Anne Marie and I watch as Thomas's sow draws in the contents of the semen bag. It happens so fast you expect to hear that sucking noise from a straw at the bottom of a milkshake. The sow appears calm and content, but she does not appear to inhabit a frenzied, ecstatic physical state. I ask Anne Marie whether this pig is having an orgasm.

"We don't know," she answers. "And to be honest? We don't really care whether she has an orgasm. We just know these contractions seem to improve the semen transport and the fertility." I must have registered chagrin, because she was quick to add, "Me in person, I think it would be nice for her. But it doesn't improve the economy of pig production."

I watch Martin's sow carefully, to see if her expression shifts as the semen is drawn in. I cannot discern a change. Anne Marie cautions against making assumptions. Pigs can be in a great deal of pain, she says, without it registering obviously on their faces, so presumably intense pleasure might not register either—or not in a way we recognize. Animal and human faces are wired quite differently. Our mouths and lower faces are generally more mobile and expressive than those of animals. Whereas animals express emotions more with their upper face—in particular, their ears.

Few scientific studies directly address the question of animal orgasm, because most researchers, like Anne Marie, have little reason to care. One who cared was a graduate student I'll call Carl Kendall. In his master's thesis "Orgasm in Female Primates," Kendall relates that he "manually stimulated the circumclitoral area and vagina of several adult, adolescent, and juvenile chimpanzee females." What ensued? Orgasms. How did he know? He felt it. "During intravaginal stimulation,* perivaginal muscular contractions of about .8-second duration and about one second apart were palpated. The average number of digital thrusts (at an approximate rate of one to two per second) performed before the onset of the contractions was 20.3." Meaning that the chimps were having orgasms after as little as ten or fifteen seconds of thrusting. As speedy as this sounds, it's not speedy enough; male chimps ejaculate after five to seven seconds. Meaning that Kendall was most likely delivering a rare treat. By the end of the passage, his

---

*Finger-fucking.

scientific veneer had eroded somewhat: "On one occasion, . . . this female reached back to grasp the thrusting hand of the experimenter and tried to force it more deeply into her vagina."

Kendall reported that when a female colleague watched one of the "experimental sessions," she found it hard to believe that the ape was having an orgasm, because the animal's face registered so little emotion or pleasure—even while Kendall was "palpating intense vaginal contractions."

And there is your answer. Female animals can have orgasms, after very little stimulation, and without it registering on their faces.

And sometimes *with* it registering on their faces. Endocrinologist D. A. Goldfoot studied stump-tailed macaques, a primate species in which the female is occasionally observed making the same round-mouthed "ejaculation face" that the males make. (A photograph is included in the paper; picture a person blowing smoke rings.) Interestingly, the face was observed most often on females that had mounted another female and been making thrusting motions.

To be sure the stump-tails' facial expression corresponded to orgasmlike contractions—rather than being merely an imitation of male behavior—Goldfoot put a strain gauge in the uterus of an especially enthusiastic female "mounter" and then put her in an enclosure with five other females. A graph charting the force of the monkey's uterine contractions appears in the study. During the nine seconds that she wore her ejaculation face, an enormous protracted peak appears on the graph.

Alfred Kinsey brings us additional evidence of female-to-female bliss in the animal kingdom. Cows mounting and thrusting upon other cows, he writes in *Sexual Behavior in the Human Female*, will sometimes give "a sudden lunge at the peak of response . . . then drop back into inactivity as though they had experienced orgasm."* Kinsey's source for the cow "data" is our old friend Dr. Shadle, at that time a lecturer at the University of, delightfully, Buffalo.

While it is often true that people are pigs, it is never the case that pigs are people. If you really want to know how sperm make their way into a woman's uterus and whether orgasm has anything to do with it, you should probably study a woman rather than a pig or a monkey. This fact was not lost on history's gynecologists, and they have done their best, if not always their brightest. A nineteenth-century physician named Joseph Beck writes in his 1874 paper "How Do the Spermatozoa Enter the Uterus?" that inquiring medical minds have upon occasion done autopsies of women who died suddenly during sex. Beck does not say how the women died, but let us assume—or anyway hope—that they died of a heart attack or stroke brought on by an intense orgasm. As with the hamsters and the dogs and the rats, sperm were typically found to have already made their way to the woman's uterus.

Beck felt confident that some sort of uterine upsuck happened during orgasm, and that this was pulling the

*For anyone who doubts that a bisexual sexual orientation exists among farm animals, allow me to quote from p. 100 of *The Artificial Insemination of Farm Animals*. Under item 4 of the section on using teaser cows to arouse a bull before taking his seed, we read that "males are just as effective as females for providing sexual stimulation and are more effective for some bulls." Also arousing for bulls: the idea of the threesome. "Even two teasers which are ineffective when used singly may bring about stimulation when placed together."

sperm along toward the egg. The only way to know for sure, he wrote, would be to watch the cervix "during the sexual orgasm." And that is what he did. Helping him out was a thirty-two-year-old blonde with a prolapsed uterus (and, Beck adds, for no particular reason, persistent constipation and acne). In other words, this woman's cervix—the gateway to her uterus—was parked in plain view, directly inside the opening of her vagina. Conveniently, this was a woman of such "passionate nature," as she herself warned Beck, that he must be careful in his examinations. For she was "very prone . . . to have the sexual orgasm induced by a slight contact of the finger."

Beck took advantage of this rather exceptional set of circumstances. "Carefully, therefore, separating the labia with my left hand, so that the [cervix] was brought clearly into view in the sunlight, I now swept my right forefinger quickly three or four times across the space between the cervix and the pubic arch, when almost immediately the orgasm occurred, and the following is what was presented to my view: . . . Instantly that the height of excitement was at hand, the [cervix] opened itself to the extent of fully an inch, as nearly as my eye could judge, made five or six successive gasps as it were. . . ." To bolster the case for upsuck, Beck points out that the cervix reminded him "precisely" of the pendulous upper lip and round mouth of a freshwater fish called the sucker.

Beck had convinced himself that "the passage of spermatic fluid into the uterus is explained fully, satisfactorily, and in every way beyond the shadow of a doubt." Just in case, he throws in a quote from a noted peer: gynecologist Marion Sims. Sims envisioned the cervix as "an India rubber bottle slightly compressed so as to expel a portion of its contents before placing its mouth in a fluid." "Hear him!" cries Beck, adding, in an impressive display of professional upsuck, "Indeed words are powerless to express my admiration for his acuteness."

Beck carries on with corroboration from his colleagues, each of whom had stumbled onto his own version of Beck's excitable, prolapsed blonde. Either women have changed since the early 1900s, or gyno exams have. Hear this: A Dr. Wernich describes patients who are aroused by "the mere sight . . . of the preliminary preparations for an examination." Wernich, in turn, relates the experience of his colleague Dr. Litzmann: "I myself recently had occasion to observe, while examining a young and very excitable female, that the uterus suddenly took on a vertical position and sank down into the cavity of the pelvis; that the mouth of the womb became . . . rounded, softer and more easily entered by the exploring finger; and that at the same time the high grade of sexual excitement under which the patient was laboring, manifested itself in her hurried respiration and tremulous voice."

Then there was Dr. Talmey,* who, writing in a 1917 issue of the *New York Medical Journal*, relates the tale of a patient who suddenly sat up during an exam, exclaimed, "Doctor, what are you doing?" looked the examiner over "from head to foot," smiled and said, "Oh, it is all right"

---

*Not to be trusted. Talmey also makes the claim that a woman deprived of semen—either because she's not having sex or her man is using a condom—will suffer a "veritable thirst for sperma" and "eventually become a nervous wreck." Ridiculous. Or is it? In 2002, a team of SUNY Albany psychologists published a paper called "Does Semen Have Antidepressant Properties?" Of 293 female college students who took a survey, those having sex without condoms were less depressed than condom users and women not having any sex. How depressed the women were was not linked to whether or not they were in a relationship or whether they were on the pill. Reactions to the paper, principal investigator Gordon Gallup, Jr., told me, have been "largely skeptical."

and lay back down again. The reason, she later confesses, is that she "experienced an orgasm during the examination of the same quality as in erotic congress and hence thought she was being abused."

I described these men's findings to my own gynecologist, Mindy Goldman, an associate clinical professor of obstetrics and gynecology at the University of California, San Francisco. "Interesting . . ." said Goldman in an email reply, adding that she had not, in thirteen years, encountered a woman who responded this way during an exam. The cervix, she pointed out, is relatively insensitive to touch—so much so that biopsies are often done without anesthesia. In a small investigation by Alfred Kinsey, 95 percent of the women whose cervix was stroked with a Q-tip or metal probe were unable to feel it.

Masters and Johnson, for their part, were vigorous upsuck skeptics. In *Human Sexual Response*, they point out that the uterine contractions of orgasm are "expulsive, not sucking or ingestive in character." They originate at the far end of the uterus and make their way toward the cervix, just as they do when they help expel a baby or a placenta. The pair obtained graphic evidence of these expulsive contractions while undertaking a study of masturbation as self-medication for menstrual cramps and backache. Fifty menstruating women masturbated with a wide-open speculum in place, such that it provided the researchers with an unobstructed view of the cervix. "During the terminal stages of orgasmic experience . . . menstrual fluid could be observed spurting from the external cervical [opening] under pressure. In many instances, the pressure was so great that initial portions of the menstrual fluid actually were expelled from the vaginal barrel without contacting either blade of the speculum." I do so hope they wore lab glasses.

Critics of this work point out that uterine contractions—minor peristaltic versions of which are happening
all the time, not just during orgasms—have been shown to
reverse direction over the course of a woman's menstrual
cycle. Around ovulation, when a woman is most fertile,
they pull material in toward the uterus; during menstrua-
tion they expel it. (The reproductive system is smarter
than you think, and utterly goal-directed. Not only do sex
hormones orchestrate the direction of your uterine contractions, they dilate only the fallopian tube that contains
the ovum, so that more semen ends up on that side. They
even oversee the quantity and viscosity of your discharge.
Around ovulation, cervical mucus becomes more abundant
and takes on the stringy consistency of an egg white, providing sperm with a sort of rope ladder into the uterus.)

In a follow-up study, Masters and Johnson outfitted
a squadron of masturbating women, six in all, with cervical caps that had been filled with a substance similar to
semen: same surface tension, same density. The substance
was radiopaque, meaning that it would show up on X-rays.
So if indeed it were sucked into the uterus during the
women's orgasms, the researchers would be able to document it. X-rays were taken during and again ten minutes
after orgasm. In the end, there was no evidence of even
"the slightest sucking effect." Here again, there are critics
of this work. Some say that the cap would have made suction impossible.

Masters and Johnson had other reasons to be dubious. Their internal home movies had shown no evidence
of gasping, sucking, or otherwise fish-mouthed cervixes.
What they had shown was a bizarre cervical by-product of
late-stage arousal called "vaginal tenting," wherein the cervix begins to pull away from the other side of the vagina,
creating a peaked space—or "seminal reservoir"—akin to

the upper reaches of a circus tent. (One theory is that this tenting evolved because it improves the odds of conception by creating a pocket to hold the sperm at the upper end of the vagina, preventing what one team of researchers—sounding more like economists than sexologists—dubbed "flowback losses." Of course, if the woman isn't on her back, the reservoir would be upside down.) But Masters and Johnson make the point that the tenting cervix is pulled away and out of contact with the semen. And if the cervix isn't in contact with the semen, it hardly matters whether it's sucking or not. The straw isn't even in the pop.

And possibly, if conception is the goal, you don't want it to be. Sex physiologist Roy Levin points out that sperm straight out of the penis are not yet up to the job of fertilizing an egg. They need time to capacitate. If all the sperm were immediately sucked up into the uterus, you'd be presenting the egg with duds. "Arguably, on this basis," Levin writes, "coitally induced orgasms and reproductive fitness could be *in*compatible."

time to check in with some modern fertility experts. See what they have to say on the subject of orgasm and sperm transport. The American Society for Reproductive Medicine supplied, as a spokesman on the topic, an adjunct professor of obstetrics and gynecology at the University of California, San Francisco. "My whole professional career for the last thirty years has been just infertility," said Bob Nachtigall. "And I have never had a patient ask me about that."

And if one did, what would he say?

"To the degree that orgasm sets up uterine contractions, you could argue that it could potentially be useful in sperm transport."

"You could, but you won't?"

He sighed. "I think by now you know how science is. You think you know a lot until you start to ask some really basic questions, and you realize you know nothing. I know a lot about artificial insemination, but I have no idea about the answer to your very simple question."

So why hasn't anyone done a study comparing women's conception rates following sex with and without orgasm? Because it wouldn't be simple, Nachtigall said. "You'd need sperm counts on all the men. You'd need physiological proof of whether or not the woman had an orgasm. And because we know it's possible to get pregnant without having an orgasm, you'd need a very large subject pool to prove that it wasn't just random chance."

There is perhaps another reason this study will not get done. "By the time a couple gets to an infertility doctor," said Nachtigall, "their sex life is shot. The intimate, fun, stress-reducing aspect of it is long gone. It's *work*. I for one would not want to interject orgasm into the strategy plan for infertility. If we were even to give them the faintest whiff of 'Gee, if you had more orgasms. . . .'" Nachtigall said that infertility is often perceived as a challenge to one's sexual identity. "The implication is always, 'Oh, you're not doing it right.' Couples really, really hate that. It's a very sensitive area."★

a s far back as Dr. Beck and as recently as Masters and Johnson, there was no such thing as magnetic resonance imaging (MRI), and medical ultrasound was in its

---

★It's especially sensitive when you're an infertility patient who works as a professional inseminator. Anne Marie Hedeboe related the story of one of her colleagues who was traveling overseas to adopt a child. "You can imagine, he gets a lot of teasing."

infancy. I found myself wondering whether modern-day high-tech imaging techniques have shed any light on the secret processes of fertility—or on anything else about sex. And, if so, about how you convince someone to have sex in front of an ultrasound technician or inside an MRI tube.

5

What's Going On in There?

The Diverting World of Coital Imaging

hough two will lie down, the bed is a single. It is a hospital bed, but more enticing than most. The bottom sheet is crisp and smoothed, and the bedclothes have been turned down invitingly, at an angle. Two sets of towels and hospital johnnies are stacked neatly at the foot. The effect is not unlike that of the convict's last meal: a weak bid for normalcy and decency in what will shortly be a highly abnormal and, to some people's minds, indecent scenario.

For the first time ever—after hours and behind locked doors in an exam room in the Diagnostic Testing Unit of London's Heart Hospital—a scientist is attempting to capture three-dimensional moving-picture (or "4-D," time being the fourth dimension) ultrasound footage of human genitalia in the act of sexual congress. Jing Deng, a senior lecturer in medical physics at University College, London, Medical School, has made his name developing a new technique for viewing anatomical structures in motion. His Web site includes fairly astonishing 4-D ultrasound footage of, for instance, beating hearts. This kind of imaging gives surgeons a preview of the structure they'll be operating on, in motion and from any perspective. It allows them to see precisely what the problem is and how they might best approach it, long before picking up the scalpel. Deng's paper on the imaging of the musculature of a pair of puckering lips—undertaken to help a plastic surgeon hone his strategy in a cleft palate follow-up operation—made it into the *Lancet*, more or less the *New Yorker* of medical journals.

In his most recent paper, Deng filmed a 4-D "erecting penis." With genital imaging, the hope is that the technology might afford better diagnostics and more detailed insights into surgical options for patients with vascular or structural abnormalities, such as Peyronie's disease, in which scar tissue in the erectile chamber on one side of the penis causes painful, crooked erections.

. Deng is the first to gather moving images of internal sexual anatomy, but not the first to use ultrasound to study sex. In 2007, a team of French researchers scrutinized images of a woman's clitoris as she contracted a certain pelvic floor muscle (the levator ani). They noticed that this contraction—which other researchers have shown to be triggered reflexively during penetration—pulls the clitoris closer to the front wall of the vagina. "This could explain the particular sensitivity of the G spot and its role in orgasm," the team wrote. And without ultrasound, no one would ever have known.

In the penis paper, Deng mentioned the possibility of one day soon capturing an ultrasound sequence of real-time two-party human coitus. Though the first few scans would be dry runs to see if the technique works and whether it reveals anything new about coital biomechanics, Deng envisions the scan as a potentially useful diagnostic tool—for instance, in teasing apart the possible causes of dyspareunia (painful intercourse).

I sent Dr. Deng an email asking permission to come to London to observe the first scan. He wrote back immediately.

> Dear Ms. Roach, Many thanks for your interest in our research. You are welcome to interview me in London. . . . However, to arrange a new in-action would be very difficult, mainly due to the difficulty in recruiting volunteers. If your organization is able to recruit brave couple(s) for an intimate (but noninvasive) study, I would be happy to arrange and perform one.

My organization gave some thought to this. What couple would do this? More direly, who wanted to pay the three or four thousand dollars it would cost to fly them both to London and put them up in a nice hotel? My organization balked. It called its husband.

"You know how you were saying you haven't been to Europe in twenty-five years?"

Ed was wary. It was not all that long ago that his agreeable nature, combined with a touching and foolhardy inclination to help his wife with her reporting, landed him in a Mars and Venus relationship seminar that involved talking to strangers about his "love needs."

I pushed onward. "What if I offered you an all-expense-paid trip to London?"

Ed sensibly replied that he would want to know what the catch was.

I read aloud to him from an information sheet that Dr. Deng had emailed. "*Dynamic 3D ultrasound imaging is a non-invasive and harmless technique which has been used for clinical imaging of activities of unborn babies. We are investigating whether this technique can be used to reveal more information on how various body parts work during various activities. . . ."*

Ed wanted to know *which* various body parts. I skipped ahead on the information sheet. For instance, I skipped the paragraph that says, *"For a dry penile scan, a volunteer is asked to lie on the bed facing down, and place his penis through a hole in the bed into an artificial vagina. The 'vagina' is made of (harmless) starch jelly."*

"Um, let's see," I said. "'Volunteers will be asked to place their body parts of interest . . .' So it's basically just the body parts of interest. We could take a day trip to Stonehenge, see a couple plays. Jeremy Irons is in something, he has a big beard now."

Ed doesn't care about Stonehenge or Jeremy Irons. But he agreed anyway.

It is a simple and noble goal: *To reveal more information on how various body parts work during various activities.* In the case of the activity known as sexual intercourse, it is an undertaking that began five centuries ago. In 1493, the artist, inventor, and anatomist Leonardo da Vinci drew a series of sketches of the commingled nether regions of a man and a woman. Known as "the coition figures," these cross-sectional cutaways were meant to reveal the arrangement of the reproductive organs during sex.*

---

*It's possible they also revealed Leonardo's distaste for intercourse. Anatomist A. G. Morris describes one of the coition figures as a "quickly scrawled illustration . . . on the corner of a page filled with mechanical drawings of cranes, pulleys and levers." It was as though Leonardo had set out to work on sex but got distracted by engineering—a scenario that no doubt plays itself out in reverse in the notebooks of countless college engineering students. "Copulation," Leonardo wrote, "is awkward and disgusting." He is said to have never bedded a woman.

Leonardo* learned about anatomy by studying cadavers. When I came across the coition figures, I assumed— erroneously, *ludicrously*, you might even say—that Leonardo had managed to wrestle two cadavers into the missionary position, and then cleave the joined couple lengthwise. The assumption wasn't entirely far-fetched; the anatomist spoke of dissecting hanged murderers (the only bodies made available for dissection), whose corpses, owing to the hanging, often, as Leonardo wrote it, "have this member rigid."

But the coition figures were not drawn from cadavers. In the frankly titled journal paper "On the Sexual Intercourse Drawings of Leonardo da Vinci," South African anatomist A. G. Morris points out that Leonardo's dissecting years commenced some twenty years *after* the sex figures were drawn. Leonardo was working from a series of ancient, and anatomically fanciful, Greek and Arabic medical texts. If he'd been working from a careful dissection of cadaver loins, presumably Leonardo would not have left out the ovaries and the prostate. Nor would he have drawn a tube connecting the woman's womb and breast, reflecting the medieval belief that breast milk was formed from (gack!) diverted menstrual blood. Not surprisingly, the mechanics of the act are also misportrayed. The penises in some of Leonardo's sketches have pushed clear through the cervix, which has opened up, Pac-Man–like, to accommodate them.

The next artist-cum-scientist to apply his motley talents to sex was the gynecologist Robert Latou Dickinson.

---

*Scholars (and rubes like me) use Leonardo rather than da Vinci as a shortened print reference because da Vinci isn't his surname. Da Vinci refers to where he was from: "of Vinci," a town in Tuscany. Somewhere along the line, these place references began to be treated as regular surnames, sparing us from having to similarly exalt Leonardo DiCaprio.

From the 1890s to the 1930s, Dickinson gathered data for his eclectic and groundbreaking *Atlas of Human Sex Anatomy*. He did make use of cadavers—preserved parts, not whole bodies—but he regarded them askance, for, as he put it, "there is marked contrast . . . between the quick and the dead. . . . The post-mortem uterus droops, the scrotum sags, the anus gapes widely." Whenever he could, Dickinson took his data from the living. He made tracings of wombs from X-rays and crafted, over the years, 102 plaster casts of patients' hymens, vulvas, and vaginas in all their various forms and states.*

It would seem, from looking through Dickinson's books, that there was no line of inquiry or request that the man shied away from. Including, "Would it be okay if I slid this test tube up your vagina?" The test tube was sunk repeatedly at differing angles, yielding the surprisingly varied profiles of women's vaginal cavities. Figure 57 in his *Atlas* shows us three life-size outlines, one inside the other, in the manner of those humane society logos with the bird silhouette inside the cat silhouette inside the dog. In place of pets, we have "long post-menopause" inside "virgin" inside "vigorous and varied coitus." The last one is as big as a grown-up's mitten.

The test tube also served as Dickinson's solution to the challenge of drawing the genitals during the actual act of

---

*The indefatigable Arlene Shaner, at the New York Academy of Medicine, tracked down a portion of this collection for me. Photos arrived in my email box ("I hope you had a lovely Thanksgiving. I am attaching some Dickinson vaginas for you") and knocked me flat. The castings are presented in beautiful arched alcoves, like bas-reliefs of saints on the walls of a chapel. Next time you're in Brooklyn, stop by the SUNY Downstate archives and ask to see the Dickinson vulvas. Or, you know, don't.

sex. He assumed that the glass tube would follow the trajectory of the penises that had come before it. Thus, he could tell the angle of the penis relative to the woman's various reproductive organs and, by shining a light into the test tube and peering down the length of it, he could see where the tip made contact during sex. Figure 91 shows a cutaway of a vagina with the tube inside and the words "Test-Tube of 1¾ Inch Demonstrates Penis Action" written along its side like a slogan on a hardware store yardstick.

Dickinson was eager to rebut claims being made that a man's penis, during sex, drives straight on into the cervix and that the two interlock, as Leonardo had drawn. Among Dickinson's papers is a manuscript of a 1931 article by Marie Carmichael Stopes, entitled "Coital Interlocking." Stopes, best known for founding Britain's first family planning clinic, was a bit out of her element here.* She had no M.D. She had trained as a paleobotanist, not as an anatomist. Nonetheless, Stopes claimed to have observed forty-eight examples of the cervix opening wide and then "closing round the glans penis as a result of the stress of sexual excitation." The first case, she writes, was a "direct observation in myself." Stopes's claims were, to use her terminology, "poo-poohed" by gynecologists—including Dickinson, who penciled exclamation marks up and down the margins of his copy of her paper. Still, you have to marvel at a woman who, in the 1920s, in the name of science, was masturbating with a speculum in place and a mirror between her legs.

---

*Something of a theme for Marie Carmichael Stopes. Her popular and controversial sex manual *Married Love* was written while she was still a virgin. Either she got some things wrong, or she failed to follow her own advice: Stopes's 1911 marriage was annulled, unconsummated, three years later.

Dickinson, wielding his test-tube spyglass, found that interlocking—or at least its precursor, head-on penis-cervix contact—was a far rarer occurrence than Stopes had suggested. It seemed to be limited to women whose cervix and uterus were abnormally positioned and to those in the "knee-chest" posture.*

Dickinson's discovery landed loudly in the fledgling field of fertility. Many physicians at that time were preaching that the failure of a couple to achieve a good interlock resulted in infertility. Now they'd need to look elsewhere for the culprit. Developers of early birth control techniques paid heed as well. Stopes had claimed that because the penis docked inside the cervix and thus delivered its payload directly into the uterus, a dose of spermicide in the vaginal cavity was of no use. This was wrong. (It was of *limited* use, but useful nonetheless.)

If the cervix truly did open wide and then clamp down on the tip of the penis, this could spell trouble for condom users. Indeed, Stopes cites a letter to the editors of the *British Medical Journal,* in which a Dr. Maurice B. Jay was called upon to see a woman with a unique and troubling situation down under. She explained that during sex earlier that day, something inside her had grabbed and torn away a piece of her husband's condom and then gripped it so firmly that she couldn't pull it free. Upon examining the woman, Dr. Jay determined that the rottweiler inside was her own cervix. Jay writes in his letter that he found two inches of

---

*You are perhaps envisioning, as I did, a woman with her knees pulled up to her chest. It wasn't until I reached Dickinson's Figure 155 that I realized this wasn't what he meant. The drawing shows a woman on her knees in a modified doggy-style posture, her chest and one ear pressed to the floor, as though listening for hoofbeats.

the sheath "firmly fixed in the cervical canal," adding that "some force was required to pull it out."

A letter published the following week questioned Dr. Jay's conclusions and nominated muscle spasms in the vagina as the mystery condom ripper. Either Dr. Jay was in need of a gynecological refresher course, or the woman did indeed have a grasping cervix. My conclusion, a conclusion you will encounter many times in the course of these pages, is that the sexual anatomy and responses of the human female are as uniform and predictable as the weather.

It would be eighty years before someone took the coital-imaging baton from Dickinson and ran with it. In 1991, Dutch physiologist Pek van Andel was looking at a cross-sectional MRI of a professional singer's mouth and throat as she put up with what must surely have been the worst acoustics of her career and sang "aaaah" inside an MRI tube. The image, van Andel said, brought Leonardo's sex figures to mind, and he found himself wondering whether it would be possible to "take such an image of human coitus."

Van Andel teamed up with gynecologist Willibrord Weijmar Schultz, radiologist Eduard Mooyaart, and business anthropologist* Ida Sabelis. Dr. Sabelis's anthropological role in the project is not explained in the paper; however, as you will see from her account of the project, no one can accuse her of being a lame duck in the proceedings:

---

*A business anthropologist is someone who, among other things, helps corporations avoid cross-cultural misunderstandings. For example, the one that led PepsiCo to run an ad in the Chinese Reader's Digest that said, "Pepsi brings your ancestors back from the grave!"—rather than the intended "Come alive with Pepsi!"

In the autumn of 1991, Pek phoned my partner Jupp. Whenever he does that, he mostly has something special on his mind. The point was to visualize with a modern scan how it really shows when a man and a woman are making love. . . . Pek suggested it should be just something for us, [because] we are slim, and because of our background as acrobats. . . .

After some shifting of dates, 24 of October was fixed as the day. I was worried, now it was really going to happen. . . . What should colleagues say? And neighbors, friends, family? . . . How shall it be in such a sterile white tube? . . . What shall we do when one of us shall get not any sexual arousal in that thing? . . .

Willibrord was waiting for us in the hall. . . . Eduard has tuned the machinery. The window between [the MRI tube] and control-panel is covered with large blue pieces. But how can someone starts such a thing? Again, as in the first conversation with Willibrord, with a talk about the weather. Pek . . . is telling us about an article he's going to write. . . . Another cup of coffee and then I say, "Jupp, shall we do something . . ."

We undress ourselves, lying down on the sledge-bed and are slided in by Eduard. We are lying on our side and facing each other. . . . Confined by the space we make the best of it. . . . The first shots are taken: "Now lay down very still and holding your breath during the shot!" . . . We are giggling a lot, because . . . an erection . . . simply sinks down like an arrow when you have to hold your breathe during many seconds. . . .

It's becoming pleasantly warm in the tube and

we truly succeed in enjoying each other from time to time in a familiar way. When the microphone is telling us that we may come—insofar possible—we burst out into a roar of laughter and some moments later we do what is the purpose. . . . Sniggering we lay down a while before we announce that we just now like to go out. Like buns which are pushed from the oven we are coming outside.

Enthusiasm everywhere, it works and, we get dressed quickly to look at the shots in the control room. Of course some are blurred because of movement. But some other are of an amazing beauty: that we are! Not so much a passport photo for daily use, but surely a shot that shows so much that it makes me speechless. There, it's my womb and surely, on that place is Jupp, naturally in a way as I know from my own sensation: below the cervix. Two days later I'm feeling a kind of pride: we tried and succeeded!

It almost didn't happen. Lacking funds, the team was initially forced to use the MRI at their local hospital, part of Groningen University. This was an older model that required the couple to hold perfectly still for almost a minute, which is how snails but not people have sex. All but one man lost his erection. Only Ida and Jupp were able to "perform coitus adequately" in the MRI tube, which was a mere twenty inches high. Schultz speculates that their success had to do with their experience as amateur street acrobats: They were accustomed to performance anxiety and odd physical feats.

Eventually, the team secured permission from a better-equipped hospital whose MRI required scanees to hold still for only twelve seconds. Alas, it was around that time that a Dutch tabloid got wind of the project. The paper ran a trumped-up story quoting patients with life-threatening conditions who claimed they were having to wait for their MRIs because creepy sex researchers were tying up the machines. Shortly thereafter came the letter from the hospital director, rescinding his welcome.

Fortuitously, Schultz's local hospital had by now upgraded to an MRI with the speedier exposure time, and the team moved their base of operations back to Groningen. But even with the truncated hold time, the men's erections wilted. The project was shelved for another six years, until a "godsend," as Schultz put it, arrived on the scene: Viagra. At last, in 1998, two more couples joined Ida and Jupp in the 20-Inch High Club, and the prestigious *British Medical Journal* published the team's paper.[*]

Aside from the intriguing link between street acrobatics and erectile function, what has mankind gained from Jupp and Ida? Mankind has gained a tremendous fudge factor should mankind wish to boast about the length of its penis. Before Schultz's MRIs, few had realized how much of the penis lies hidden below the surface of the skin. The "root" is nearly two thirds again the length of the "pendulous part." So if your erection is, say, six inches long, go ahead and say it's ten. I'll back you up.

At the very least, the paper laid to rest the hokum about penises routinely interlocking with cervixes. Also, we

---

[*]"Magnetic Resonance Imaging of Male and Female Genitals During Coitus and Female Sexual Arousal" won the 2000 Ig Nobel Prize in medicine. (The annual Ig Nobel Prize is a parody of the Nobel Prize.) The award afforded Schultz's team, if nothing else, the opportunity to hobnob with the Scottish emergency room doctors whose paper on toilet-inflicted buttock injuries—"The Collapse of Toilets in Glasgow"—took that year's Ig Nobel in public health.

learned that the penis—root and stalk together—"has the shape of a boomerang" during intercourse. (Leonardo had drawn it stick-straight.) But not its precise dynamics. If you hurl an uprooted penis into the air, it will not come back to you. It will most likely, and who can blame it, want nothing to do with you.

By far the most jaw-dropping sexual discovery to come to us courtesy of real-time genital scanning is set forth in Israel Meizner's "Sonographic Observation of In Utero Fetal 'Masturbation,'" a letter to the editor published in the *Journal of Ultrasound in Medicine*. Still images, two of which accompany Meizner's letter, detail a fetus, seven months old. The first shows the teensy hand poised for action. The second shows the fetus a moment later, "grasping his penis in a fashion resembling masturbation movements." This went on for some fifteen minutes, during which time Meizner stayed tuned but did not document an in utero fetal orgasm.*

b lessedly, the ultrasound department is running behind. Ed and I have a half-hour reprieve while the day's last patients are scanned. We wander up and down the corridor. At one end is a door with a sign that reads DISCHARGE LOUNGE. "Ew," says Ed. We find a café and order tea. Ed stares at his shoes. He is concerned about his ability to, as Schultz put it, perform adequately. He has taken a "godsend," however, so he'll likely manage fine.

---

*The earliest orgasm on scientific record is that of a three-year-old girl whose mother spied on her and reported her behavior to Alfred Kinsey. Kinsey duly and unwisely—this was 1953—included a detailed description of it in *Sexual Behavior in the Human Female*. Though he had no dealings with this child or others described in his book, he was accused of pederasty, a taint that dogged him for years.

"Here we go," Ed says grimly. Dr. Deng walks toward us. He wears khaki trousers and a white lab coat. His age is hard to guess. His hair, though graying, spikes youthfully in all the right places. Though he moved to London ten years ago, he speaks English cautiously and with few decorative touches. An occasional "Brilliant" or "Cheers" is the only trace of England in his words. Nuances of humor, like sarcasm, seem to elude him, or maybe he is just preoccupied with his tasks. Dr. Deng shows us where the changing room is.

"Regarding the position," he says when we return in our johnny tops. He wants us on our sides, spoons-style. (This was explained, sort of, in the instruction sheet: *We will ask the penis to be inserted into the vagina from his partner's back.*) "I think facing the wall is better," says Dr. Deng. As opposed to facing him. "That will be more romantic," he adds. On the wall, someone has hung a painting of a hillside harbor town. As though by looking at it we could convince ourselves that we were off on the Amalfi Coast—or, just as good, that Dr. Deng was. "And I will switch off the lights."

"Where are the candles and soft music?" says Ed.

"Oh, I am sorry," says Dr. Deng, straight-faced, chagrined. Then he brightens. "I can turn on my laptop. I have the soundtrack to *Les Miz.*" His efforts are sweet though pointless. There is no way to make this situation romantic, normal, *sexual.* It feels like a medical procedure, something to be got through.

Dr. Deng goes next door and returns with a 9-by-11 envelope and hands it to Ed. Inside is a copy of a U.K. version of *Maxim.* "This is very erotic," he assures Ed. The implication being, I suppose, that the sight of one's wife in a baggy knee-length hospital johnny and threadbare socks is not.

There comes a moment in cheesy horror films when a man with evil intent reaches up and bolts a door. This is the audience's cue to fear for the heroes. Fear for us. Dr. Deng has pressed the doorknob lock. I'm running my sentences together. "That's some fancy machine you've got, how did you get interested in radiology, is there a good pub nearby, we're going to need it."

Dr. Deng never tells us to lie down, but it seems that that is what must happen.

Ed is pretending to be absorbed by his magazine. I nudge him. "Jupp, shall we do something?"

We get into position while Dr. Deng applies ultrasound gel to the end of the ultrasound wand. The gel conducts ultrasound waves better than air does. Ultrasound gel looks and feels (and works) like the product euphemistically known as personal lubricant.

Dr. Deng starts by taking some still images. He reaches across Ed to hold the ultrasound wand to my belly. His arm rests on Ed's hip, a curiously intimate touch in an encounter otherwise strangely devoid of intimacy. For the still images, we must hold still for several seconds, like Victorians posing for a tintype, only not like Victorians posing for a tintype.

"Now please make some sort of movement," says Dr. Deng. And then, in case it's not clear, in case Ed might be contemplating flapping an elbow or saluting the flag, he adds, "in and out."

Dr. Deng says he's pleased with the result. "It's actually much clearer than I thought it would be. It's very— Hm. Can you just hold there, for a while? We saved too many data." Dr. Deng needs to reboot. Fortunately, it takes only a few seconds, sparing Ed the necessity of also rebooting.

Ed keeps up an idle, disaffected rhythm. He and Dr.

Deng chat about their children. I'm taking notes. Or half of me is. I feel like a secretary in a ribald French comedy, sitting calmly at her desk, taking a letter, while the mailroom guy hides in the footwell with his face between her legs.

"You look so young to have a fifteen-year-old," Ed is saying. "How old are you?"

"I'm forty-five in August."

"And the little one? How old?"

"Just two and a half. You can ejaculate now."

As far as I'm concerned, the only downside to direct imaging of real people having sex is that there will no longer be call for researchers to go out and buy dildos and pliable plastic vaginas from California Exotic Novelties and then bring them back to the lab and make them have sex together. In 2003, a team of evolutionary behaviorists at the State University of New York at Albany published a paper called "The Human Penis as a Semen Displacement Device." They theorized that man evolved a penis with a ridged glans in order to scoop out the competitors' semen before depositing his own. (The single gal of prehistory must have been fantastically promiscuous.) This would fit in with the little-known fact that the last portion of a man's ejaculate contains a natural spermicide—not intended to kill his own soldiers, obviously, but to annihilate the seed of any who come after him.

You can't buy simulated human semen from California Exotic Novelties, and so the Albany team concocted their own. Several recipes were tried and "judged by three sexually experienced males." Here is the winning recipe chosen by the judges:

Human Semen

   7 milliliters room-temperature water
   7.16 grams cornstarch

Mix ingredients together. Stir for five minutes.

Yield: one ejaculate

The simulated semen was ejaculated, via syringe, into the vagina, which had been anointed with lubricant (also from California Exotic Novelties). With a video camera rolling, three different phalluses—including a Control Phallus with no ridge—were inserted and withdrawn. To see how much competitor semen each phallus had managed to scoop out, the vagina* was weighed before and after. The results backed the team's theory: Both of the lifelike phalluses (or "dongs," to CEN shoppers) displaced 91 percent of the semen, while the unridged control dong left 65 percent of it behind.

   The experiment went on for six more pages, but to be honest, they lost me at:

$$\frac{\text{(weight of vagina with semen } - \text{ weight of vagina} \atop \text{following insertion and removal of phallus)}}{\text{(weight of vagina with semen } - \text{ weight of empty vagina)}} \times 100$$

   To my mind, what happened in Dr. Deng's exam room bore no more relation to sex than a smile held for a camera does to the real thing. It was perfunctory, self-conscious,

---

*The California Exotic Novelties Web site features five pages of simulated vaginas, many of them crafted from molds of actual porn star orifices. Meaning that it's altogether possible that, say, Alexis Amore made an appearance in the pages of *Evolution and Human Behavior*.

distracted sex. Other than the parts involved, it bore very little resemblance to what goes on between my husband and myself when there's not a strange man on one side of us and an ultrasound wand on the other. Though they no doubt have their uses, ultrasound movies are a superficial rendering of the complex and varied body-mind meld that we call sex. Sex is far more than the sum of its moving parts.

But you can't altogether dismiss the parts. If the parts don't work properly, the sum is moot. For some 18 million American men, they don't work as they ought to or, at least, as they once did. Next up: the occasionally noble, sporadically ghastly, ever-surreal world of erectile science.

6

The Taiwanese Fix and the Penile Pricking Ring

Creative Approaches to Impotence

a man having penis surgery is the opposite of a man in a fig leaf. He is concealed face-to-feet in surgical sheets, with only his penis on view. It appears in a small, square cutout in the fabric, spotlit by surgical lamps. To lie completely naked would preserve more modesty, for then the onlooker's gaze is bound to stray. There are moles and chest hair to look at, knees, nipples, Adam's apples. This way, all eyes stay on the organ.

But a man, even an impotent man, needn't feel self-conscious under the gaze of Geng-Long Hsu. Dr. Hsu, who practices in his native country, Taiwan, has been a urological surgeon for twenty-one years. Whatever one might come here for, it is safe to assume Dr. Hsu has seen worse. He has seen smaller, crookeder, pinker, limper. He has seen penises with implants poking through their tips like collar stays. One day this year, he repaired a penis that had ruptured during a performance of jui yang shen gong, an obscure martial art.★ "He tried to lift one hundred kilograms with his penis!" Dr. Hsu exclaimed yesterday while we rode the elevator to the lobby.

Dr. Hsu runs the Microsurgical Potency Reconstruction and Research Center in Taipei, where he has been researching and honing a surgical treatment for impotence. The operation, which involves tying off and removing some of the veins of the penis, has fallen out of favor elsewhere in the urological community, but Dr. Hsu believes that if it is done correctly and thoroughly, it can help up to 90 percent of men with erectile dysfunction (known among urologists as ED, to the minor chagrin of Eds the world over).

The dysfunctional penis in the spotlight this morning has been so for the last eight of its forty-seven years. The patient has tried Viagra, with limited success. The organ sits alone on its little skin stage, looking vulnerable. I find myself feeling nervous for it, as one might for a fifth-grader before a solo recital.

Once the anesthesia takes effect, Dr. Hsu will begin "degloving" the organ. The verb "skinning" would get the idea across more efficiently, but it is more pleasant, I suppose, to picture an aristocrat gently loosening the fingers of his opera gloves.

Dr. Hsu makes a cut in the flesh just above the penis.

★The world record belongs to Grandmaster Tu Jin-Sheng, who, along with two fellow practitioners, pulled a flatbed delivery truck across a Taipei parking lot with their penises in October 2000. For those interested in learning the art, or perhaps starting their own penile delivery service, MartialArtsMart.com sells Tu's video *Iron Crotch*.

I'm a little confused because the *Taipei Times* referred to this technique by the name yin diao gung, or "genitals hanging kung fu"—the fifth of the Nine Mysterious Kung Fus. Mysterious Kung Fu No. Six entails "drawing water back through the urinary tract and into the bladder by use of a drinking straw," which does not ease my confusion.

He slips his blade into the wound, and slides it underneath the skin.

"Remember the night market?" he says. Last night, Dr. Hsu took me to Huahsi Street, a market known for the lurid shows put on by its snake-medicine vendors. Unlike snake shows in Morocco or India, Taiwanese snakes are not charmed. Pretty much the opposite of charmed. They are skinned alive, bled, and made into stews.

Dr. Hsu works his scalpel down the shaft of the penis, detaching the skin from the pulpy pinkness underneath. "Very like the night market!"

Dr. Hsu speaks English with great enthusiasm and fitful syntax. This is occasionally frustrating but mostly just endearing. Yesterday he took me and his business associate Alice Wen on a tour of Taipei city sights. "Okay!" he exclaimed as we piled out of the car. "Let's experience!" He is patient, polite, and unwaveringly generous. As we set off into the crippling midday heat, he doled out sun visors and baseball caps, taking the one no one picked for himself: a China Youth Camps cap with a pink bunny motif.

By his own description, Hsu is something of a "queer bird." He once inserted a Foley catheter into his own urethra, "just to feel what patients feel." He self-medicates with acupuncture, sometimes walking around his clinic with a needle protruding from the side of his head. I have heard him use a translation of one of Chiang Kai-Shek's names ("central uprightness") to make a point about penile curvature.* Today finds him wearing blue plastic wraparound

---

*A comforting word about the crooked penis. Dr. Hsu says it is rare to see one that stands perfectly straight. Actually, what he said was: "Most men are communists! Lean to the left! Second most common: bow down, like a Japanese gentleman! Number three, to the right. Four, up! Like an elephant!"

sunglasses as he operates. He explains that this is because he suffered a seizure some years back, which left him sensitive to glare. His face is still partly paralyzed. This is noticeable only when he smiles, which he does with just one side of his face, in the manner of a dramatics-club mask.

The next step in the operation Dr. Hsu has named the "inside-out maneuver." Though it is not so much "inside-out" as just "out." Using his gloved fingers, Dr. Hsu pulls the man's penis up and out of its skin, through the three-inch slit, by its midshaft. The protruding skinless portion is doubled over inchworm-style. (The skin at the head of the penis has been left attached.) I ask Dr. Hsu what this maneuver would feel like without anesthesia. His answer: "Like the way to treat a spy."

For the next three hours, Dr. Hsu isolates the veins he is after, ties off the blood supply with sutures half again as fine as a strand of his hair, and then snips the veins away. He begins with the deep dorsal vein, the fattest. Bit by bit, he frees it from its moorings. As he works, he pulls the vein taut and holds it out away from the penis, like a robin pulling on a worm.

t o understand why removing veins from a penis can help it stay erect, you need to understand how it gets that way in the first place. Erections are all about blood. Blood is the backbone of a stiff penis. Though it was a long time before anyone figured this out. In the Middle Ages, the erect male member was thought to be filled with pressurized air, a miniature skin blimp. It was Leonardo da Vinci who made the breakthrough. Cadavers available for anatomy study back then were typically those of executed murderers. Because they'd been hanged, the dead criminals had erections, and because Leonardo was dissecting them, he noticed that their penises were, in his very own words, "full of a large quantity of blood."

The blood resides in a pair of cylindrical chambers—the corpora cavernosa—which lie side by side like a diver's tanks. The chambers are filled with smooth-muscle erectile tissue, full of thousands of tiny hollow spaces, like a sponge. When the smooth-muscle tissue relaxes—which it does at the behest of an enzyme activated when the brain perceives a sexual stimulus—it expands. (Smooth muscle, unlike the striated muscles of your arms and legs, is operated by the autonomic nervous system; this is why men can't simply will themselves erect—or unerect.) The relaxation of the erectile tissue allows blood to rush in and fill out the spongy hollows. Drugs like Viagra enhance the erection process by knocking out a substance nicknamed PDE5, which inhibits smooth-muscle relaxation. They inhibit the inhibitor. (Thus, they're called PDE5 inhibitors.)

So now we have *achieved,* in the parlance of ED experts, an erection. It is a respectable achievement, but it is not enough. An erection, like a motorcycle or a lawn, must also be *maintained.* The blood that has filled the two erection chambers* must be trapped there, otherwise the erection wilts. This is tricky, as the chambers are equipped with drainage veins along their surface. What keeps the blood from leaking out via these veins? The miracle of passive venous occlusion. (Stay with me here.) These drainage veins lie outside the erection chambers but inside the stiff outer membrane (called the tunica) that protects the

*Actually, there's a third, that runs beneath these two, but it's a lesser player and we're going to ignore it. Likewise, we are going to ignore the erectile tissue in the lining of the nose—which does, very occasionally, expand when its owner is sexually aroused. It too is made erect by increased blood flow. Nasal congestion is an erection inside your nose.

erectile tissue. When the chambers expand with blood, they slam up against the tunica—which also expands, but not as much—and this pressure squeezes shut the veins caught in between. If all goes well, the blood stays trapped until a postorgasm chemical messenger tells the smooth-muscle tissue to stop relaxing.

When a man is impotent, very often it's because the erectile tissue isn't expanding as vigorously as it needs to squeeze shut the veins, and some of the blood seeps out. The result: "Like a tire! Flat!" Dr. Hsu relies on a lively repertoire of metaphors and analogies to explain the various functions and dysfunctions of the male genitalia. One particularly ambitious explanation, delivered earlier today, involved a Christmas tree and an elephant's trunk. A diver was diving into a pool, and an aircraft was taking off. I felt like ducking under the table.

The most common explanation for ED is that the erectile tissue is simply getting old. "As we age, we lose elastic fibers, we lose smooth muscle, our tissues become more rigid," explains Gerry Brock, a professor of urology at the University of Western Ontario who sits on the board of the *Journal of Andrology*. In an offshoot of aging called fibrosis, some of the muscle cells in the erectile chambers are gradually replaced by fibers of connective tissue that don't have the elasticity that youthful smooth-muscle tissue has. When erectile tissue loses its stretch, it no longer expands fully and presses hard against the walls of the tunica. Thus, the veins aren't squeezed shut, and blood leaks out. If you were to tie off and remove some of those veins, it would prevent—or at least slow—the leakage.

Because the largest of these drainage tubes, the dorsal vein, runs just under the skin along the top of the penis, it's possible to effect a crude version of Hsu's surgery by simply constricting the thing with an elasticized band or clamp.

A surprising amount of this went on in the pre-Viagra era, long before the cock ring entered modern vernacular. There are so many patents on file for erector rings* that they earn their own chapter in Hoag Levins's diverting book *American Sex Machines: The Hidden History of Sex at the U.S. Patent Office*. Levins traces the ring's evolution, from the handsome steel clamps of the circa 1900 metal-machining era clear through to a 1989 model with a hand-held remote, stopping along the way for a double-page spread of "Penile Ring Clamp Patents of the Post-WWII Years." Though very few of the early patent titles contain the words *penile* or *erection*. Many employ the unhelpful rubric "Appliance for Assisting Anatomical Organs." A 1900 patent is coyly titled "A Boon to Men." Descriptions are similarly vague. One 1897 clamp is for use on male organs that "fail to perform their office," leaving it unclear, at first glance, whether the device is intended to aid the organ or court-martial it.

Dr. Hsu was not the first surgeon to realize that limiting venous drainage could be a ticket to newfound potency.

---

*Of course, you can use any old ring as a cock ring. In China, I saw rings of animal tail skin being sold for this purpose. Robert Latou Dickinson has a note in one of his files about the eyelids of a sheep, the lashes intended as a sort of ovine French tickler. Also on record: the handle hole in the head of a sledgehammer.

Trouble is, without a release latch, you might not get it back off again. In San Francisco, cock-ring emergencies are so common that they have their own shorthand ("C-Ring") on the Fire Department teletype. The department's Heavy Rescue Squad has modified a small circular saw especially for this purpose and occasionally stages practice drills. The latter prove challenging owing to the absence of manipulable genitalia on Resusci-Andy dolls and the refusal of male staff to volunteer as mock victims.

This footnote dedicated to former HR squadder Caroline Paul, who personally liberated four penises, including that of the sledgehammer guy, who did not say thank you.

A Dr. Joe Wooten tied off a man's dorsal vein with catgut in 1902. The high risk of infection—penicillin hadn't yet been discovered—probably kept most of Wooten's colleagues from taking up the scalpel, but it might also have been Wooten's flummoxing conclusion in a *Texas Medical Journal* article from that year: "It has now been about four months since the operation, and the party reported to me . . . that he had had for the first time in nearly three years complete and satisfactory coitus and was now willing to stop trying."

It was Hsu's mentor and fellowship advisor Tom Lue,* a professor of urological surgery at the University of California, San Francisco, who refined and championed the procedure in the late eighties. Alas, a long-term cure proved elusive—the success rate dropping in one study from 62 percent after three months to 31 percent after forty-five months. Why would this happen? The logical reason, say Brock and others, is that the body tends to compensate when someone or something destroys or blocks a vein. It grows new veins, and/or the remaining ones get bigger. Lue ultimately distanced himself from the procedure, and others followed suit.

Most everyone but Dr. Hsu. In a 2005 paper published in the *Journal of Andrology*, twenty-one of Hsu's patients who underwent an early version of penile venous stripping

---

*Lue was the 1988 recipient of the Gold Cystoscope Award, which Dr. Hsu remembered as the "Golden Bladder Award." The ensuing trip to Google for verification revealed a veritable medical supply catalogue of gold-plated statuary: there is the Golden Speculum Award and the Golden Forceps Award. The American Rhinologic Society bestows a Golden Head Mirror Award. There are at least three different Golden Stethoscope Awards. The American Society of Colon and Rectal Surgeons, not getting into the spirit of it *at all*, bestows an annual Mentor's Award.

surgery in 1986 were contacted for a follow-up. They filled out the same questionnaire they had filled out before the surgery: the International Index of Erectile Function (IIEF). Their mean preoperative score had been 10 (out of a possible 25), and their follow-up score was 19. Hsu saw no evidence that the penises had grown new veins.

"See the difference?" says Dr. Hsu, who is by now closing up the incision. "Already. Look how engorged." The organ is visibly larger than it was when it walked in. Dr. Hsu says many of his patients report that their penis is mildly engorged all the time. He adds that many enjoy this, that it makes them feel more confident.

"So this guy. Now ready to make a home run. Like a baseball bat!"

Why would Dr. Hsu be able to cure so many men, when other competent urologists who undertake the procedure have seen, for the most part, only short-term benefits for their patients? "I can't for the life of me answer that question," says Gerry Brock. "I've seen Geng's surgery, and he does a good job. He's an honest guy, a great guy. But I have a hard time understanding, from a physiologic basis, how his results can be so distinctly different from those of others." One possibility is that Taiwanese patients are more polite—or more timid—than Western patients. Perhaps Dr. Hsu's patients are hesitant to report that the surgery's effects are fading.

Then again, the possibility exists that no one who does this technique is as good at it as Dr. Hsu. Few urologists seem as lovingly immersed in the anatomy of the penis as Geng-Long Hsu. He publishes papers on the tunica, the deep dorsal vein, the distal ligament. He has a standing order for "leftovers" at the dissection lab of the anatomy department of Taiwan Adventist Hospital: "Please give all the penises to me." Some years back, a lab tech threw away a box containing seventy-three penises that Dr. Hsu had collected from researchers and anatomy labs over the years and stored in a freezer. The memory pains him to this day.

The next time a cadaver becomes available, Dr. Hsu plans to make a detailed examination of the penile veins to see which ones account for what percentage of the drainage. He wants to know which are most critical to remove, and which make little difference, in terms of erectile function. Which begs the question, Can a dead man get an erection? He can. I have seen it myself, on a DVD of a previous research operation,* which Dr. Hsu sent me before I came to Taiwan. Standing in for blood was saline, being pumped in at more or less the same rate that the heart pumps in blood during a normal erection.

One day this week, Dr. Hsu and I walked to a temple high on a hill behind his apartment complex. He said he regularly walks the two-mile road to the top (stopping to pick up litter along the way), and when he gets there, he jogs up and down the flights of steps to the temple door fifteen times. He explained that because he feels an obligation to help as many men as he can, he does not want poor health to cut short his life.

Geng-Long Hsu is a man on a mission. He feels he has a cure for ED and wants it to be used throughout the world, but is, well, impotent to make it happen. Until other surgeons are able to replicate his success rates, the procedure will likely remain shelved everywhere but in his clinic.

. . .

---

*Because the man had been dead only a few hours and his face was covered by a surgical cloth, the video was no more gruesome than your average Dr. Hsu production. Also, someone had added a soothing soundtrack, the sort of innocuous instrumental one hears in shopping mall atriums.

g eng-Long Hsu is typical among modern urologists in his enthusiasm for the medical and surgical treatment of what had long been considered a psychological problem. In *The Rise of Viagra*, Meika Loe makes the case that urology pretty much stole impotence out from under psychology's nose. (Loe, a sociology professor, earned my abiding respect by waitressing undercover at Hooters as part of a graduate research project in gender studies.) From the heyday of Freud all the way through the behavior therapy era of the fifties and sixties, the causes of impotence were thought to dwell in the psyche. Penises went limp from unresolved neuroses, deep-seated anxieties, distraction, obsession. If you wanted help, you turned to a shrink.

All that began to change in 1980. Loe cites as the turning point the publication of a contentious *JAMA* article entitled "Impotence Is Not All Psychogenic," as well as the introduction of the vacuum pump and the penile implant, neither of which your therapist was likely to have on hand. The medicalization of impotence was underway.

Viagra sealed the deal. In 1998, Pfizer—with a cadre of media-savvy urologists in tow—launched a massive publicity campaign to announce an exciting new approach to impotence. Only it wasn't called impotence anymore; it was "erectile dysfunction." The stigma of the psychological had been removed. Impotence had morphed into a tidy biological problem treatable with a harmless pill. There wasn't something wrong with the man, there was something wrong with the plumbing. Pfizer craftily introduced three categories of ED: mild, moderate, severe. Heck, it now seemed, everyone has it sometimes, to some degree, even Bob Dole. No need to be embarrassed. Urologists—most of them consultants for Pfizer—began appearing on talk shows, chatting about "ED" as casually as the last guest had chatted about his wheat-free cookbook.

In truth, plenty of cases of psychologically based impotence exist, and it's relatively simple to sort out which ones they are. If a man is medically impotent—because his smooth-muscle tissue is damaged, say, or there's a problem with his nerves—then he won't get erections in his sleep. If the problem is purely psychological, he will. That is why diagnosis is sometimes done by checking for nighttime erections with gizmos like the RigiScan-Plus Rigidity Assessment System (with Self-Calibrating Penile Loops). Once upon a time, it was done by having a nurse watch your penis as you slept. The next generation of "nocturnal penile tumescence monitoring," as it is officially known, took the form of a strip of old-fashioned perforated postage stamps slipped around the organ at bedtime and either torn or not torn during the night. The advantage of the "postage stamp tumescence test"* was that it could be done in the privacy of one's home and—thankfully or disappointingly— no longer involved anyone in a nurse's uniform.

Even when a patient is young and the physical state of his erectile tissue is unlikely to be the problem, urologists are inclined to skip the RigiScan and try a Viagra-type drug. I asked Ira Sharlip, a spokesman for the American Urological Association and a clinical professor of urology at the University of California, San Francisco, why these men

*The U.S. Postal Service was aware of this practice, and even endorses it. USPS spokesman Mark Saunders points out that it was a nice, if modest, source of additional revenue for the beleaguered Postal Service. "Because I bet these people didn't use the stamps after that, for mail." Saunders urges care in choosing the stamps. "For instance, we recently introduced a Distinguished Marines stamp and a Muppets stamp," both of which would seem, well, just plain wrong. Saunders also felt that some stamps, like the Greta Garbo, might be considered cheating. I asked Saunders if the Postal Service even sells perforated stamps anymore. "Yeah," he deadpanned. "We've got a new dog stamp. It licks itself."

would be prescribed pills if their condition is likely to be psychological. "These patients get into a vicious cycle," he said, "where the anxiety over not being able to get an erection compounds the problem." A PDE5 inhibitor can help reverse the cycle. "We use it as a bridge. But at the same time, I have all of those patients, if they're willing, work with a sex therapist or a psychologist."

Meika Loe quotes the medical essayist Franz Alexander on the enduring appeal of medical—over psychological—approaches to conditions like impotence. "Alexander claimed that medicine's aversion to psycho-social factors harkened back to 'the remote days of medicine as sorcery, expelling demons from the body.' . . . Twentieth-century medicine was 'dedicated to forgetting its dark magical past.'"

That's too bad, because it was, as we're about to find out, pretty darn entertaining.

In the Middle Ages, the common assumption was that impotent men had been cursed by a demon or by a witch, acting as a sort of local proxy for the Devil. According to *Malleus Maleficarum*—a 1491 handbook of judicial proceedings against witches and methods of "curing" curses and spells—witches could induce both impotence and sterility. Some displayed a surprisingly sophisticated grasp of male anatomy. While impotence was achieved by simply "suppressing the vigor of the member," the sterility curse required the witch to "prevent the flow of the semen to the member . . . by as it were closing the seminal duct so that it does not descend to the genital vessels."

Witches with no formal training in andrology could employ a simpler, more fanciful approach. They made the man's penis disappear. Authorities quoted in *Malleus Maleficarum* disagree as to whether the organ is truly gone

∗
bonk

or the bewitched individual is simply under the sway of a perceptual illusion that causes him to believe it's gone. The author quotes an unnamed venerable Dominican father who, during a confession, hears a parishioner confide that he has "lost his member" to witchcraft. The priest relates that he asked the lad to remove his clothes, so that he might, likely story, check for the missing part.

The author of *Malleus* brings up the strange matter of penises stockpiled in birds' nests, which he presents as proof of a literal disappearance. "What, then, is to be thought of those witches who . . . sometimes collect male organs in great numbers, as many as twenty or thirty members together, and put them in a bird's nest or shut them up in a box, where they move themselves like living members, and eat oats and corn, as has been seen by many and is a matter of common report?" It's a question I cannot answer. I can only lament the long, dry journey that legal publishing has made in the centuries since 1491.

The recommended cures for impotence suggest that medieval authorities may have suspected that psychological, not supernatural, powers were more likely at play. The accursed man who seeks advice is invariably asked to give some thought to whom he believes might have bewitched him. He is then urged to "prudently approach" this person and to sit down and have a talk. "Soften her with gentle words." Whereupon the penis generally reappears. As a Plan B, the learned tome recommends that the accursed "use some violence."

Come the late 1700s, blame for impotency shifted from supernatural beings to men themselves. The year 1760 saw the publication of a slim, pernicious work of hyperbolic quackery called *Onanism; or, A Treatise upon the*

Disorders Produced by Masturbation. A shrewd blend of the clinical and the moral, it spread like a virus through the medical circles of Europe and the United States. Impotence was prime among the disorders said to be produced. Sperm-carrying semen was believed to be a vital source of life energy.* As with fossil fuel or health insurance payouts, there was thought to be a finite amount of it available; woe befall the man who wantonly squandered it. Masturbation and casual sex—particularly with "ugly" women, who sapped one's vitality faster than the handsome ones—led to all manner of bodily woes. (Spilling sperm into someone you love did *not* deplete one's vital juices because, quoting *Onanism* author Samuel Tissot, "the joy which the soul feels . . . repairs what was lost.")

Onanism and its imitators—*Excessive Venery, Masturbation, and Continence,* by American M.D. Joseph Howe, came out over a century later but is no less hysterical— had citizens worrying that masturbation could cause not only impotence, but blindness, heart trouble, insanity, stupidity, clammy hands, "suppurating pustules on the

---

*The semen-as-life-force concept dates all the way back to Pharaonic Egypt. When the creator, Atum, needed to create a pair of helper gods, he masturbated. His seed spawned Shu, the god of air and—surely something of a letdown—Tefnut, the goddess of humidity. This and other tales are detailed in a group of illustrated erotic papyruses, which I read about in a paper by urologist A. A. Shokeir ("Sexual Life in Pharaonic Egypt: Towards a Urological View"). Shokeir's paper includes a drawing of Atum masturbating, which he did by mouth and, as always back then, in profile.

The best-known erotic papyrus is the Turin Erotic Papyrus, housed just down the road from the Shroud of Turin. To avoid confusion, the papyrus is the one that shows people having sex in a chariot and, also, in "the Geb and Nut Position." Shokeir includes the latter in his paper, noting in his caption that "the man carries a sack over his shoulder and takes her from behind."

face," acrid belches, "a flow of fetid matter from the funda-
ment," tongue coatings, stooped shoulders, flabby muscles,
under-eye circles, and a "draggy" gait. It was the Victorian-
era version of the anticrack campaigns that you see today,
with their closeups of acne-blighted cheeks and discolored
teeth: vanity as a force more powerful than medicine.

Tissot took it to the extreme in his description of the
effects of "self-pollution" on a watchmaker referred to
as L.D.: "A pale and watery blood often dripped from his
nose, he drooled continually; subject to attacks of diarrhea,
he defecated in his bed without noticing it, there was a con-
stant flow of semen. . . ." *Hello, yes, this watch you sold me is
all sticky and stuff?* Impotence was almost beside the point.
Masturbate for a few months, and you'd soon be so revolt-
ing no one was going to climb in bed with you anyhow.

The cure for erectile troubles, then, was simple: Quit
masturbating. Stop wasting your vital sap. Dismayingly, this
included sap spilled involuntarily during sleep. Nocturnal
emissions had to be prevented too. Here simple willpower
wouldn't do the trick. You needed technology. You needed,
in the words of the U.S. Patent Office, a Device for Preventing or Checking Involuntary Spermatic Discharges.*

On the simple side, there was the Penile Pricking Ring.
Invented in the 1850s, this was an adjustable, expandable metal ring slipped onto the penis at bedtime. If the
sleeper's penis begins to expand, it forces the ring open
wider, exposing metal spikes that, should it expand still

---

*More often, the Patent Office opted for euphemism in the category
headings for antimasturbation contraptions. "Surgical Appliance" was
by far the most common. Also "Sanitary Appliance," as though a sneeze
guard or a mop could somehow keep wanton sexual impulses in check.

further, are pushed down into the flesh, awakening the sleeper. Later, higher-tech variations had the expandable ring hooked up to an alarm bell or—*suppurating pustules!*—a shock-producing current. One device monitored length rather than girth. A metal cap was slipped over the end of the organ, giving it somewhat the appearance of a muzzled dog snout. Attached to the cap by short chains on either side was a pair of clips. These were affixed to tufts of pubic hair. I will let James H. Bowen, the owner of U.S. Patent 397,106, describe the ensuing scenario. "When a discharge is likely to occur, the device is elevated with the organ, and the connections are drawn sufficiently taut as to pull the hair."

Many of these devices included an option for daytime use, along with a lock-and-key mechanism. For the true target customer was not the penitent masturbator, but the worried parent and, even more so, the insane asylum caretaker. The institutionalized lunatic who attempted to remove his antimasturbation device faced—in the words of Raphael Sonn, inventor of the Mechanical Penis Sheath—"great physical pain and possible mutilation." Sonn's patent reads like the instruction manual for something in the Marquis de Sade's basement, with "clamping members," "gripping elements," teeth, prongs, and hinges "of the tight-butt type."

Happily, parents of K-through-8 masturbators were encouraged to try less drastic preventive measures. Little hands were tied to headboards, and trousers fashioned without pockets. Hobbyhorses were taken away, and climbing ropes removed from school gymnasiums. One of the biggest spoilsports in the antimasturbation crusade was American physician William Robinson. His 1916 *Practical Treatise on the Causes, Symptoms, and Treatment of Sexual Impotence and Other Sexual Disorders in Men and Women* includes a long chapter on preventing the premature awakening of the sexual instinct in children. "I strongly urge parents to keep their boys away from sensuous musical comedies and obscene vaudeville acts," tutted Robinson, clearly something of the tight-butt type himself. "Many of my patients told me that their first masturbatory act took place while witnessing some musical show."

Mental masturbation was also to be discouraged. "It is very rare," wrote Robinson, "that people who devote all their time to severe intellectual work do not pay for it by sexual weakness or impotence." He goes on to describe the case of a famous mathematician who, during each attempt at intercourse, "would be disturbed by an abstruse mathematical problem and the attempt would fail." Somewhat contradictory advice, mentioned in *Masturbation: The History of a Great Terror,* comes to us courtesy of a Dr. Crommelinck, who advocated memorizing difficult passages on philosophy or history when overcome by the desire to masturbate.

Truly it seemed that any activity undertaken—sleeping, thinking, eating spiced food, taking in a matinee of *Mame*—led the heedless male down the path to self-pollution. A man couldn't even relieve himself without having to worry. Crommelinck urged gentlemen to avoid touching their genitals at all times, lest they inadvertently arouse themselves—even at the urinal. "Urinate quickly, do not shake your penis, even if it means having several drops of urine drip into your pants."

Those who could not manage to curb their impulses with philosophical tracts and antimasturbation gadgetry faced a withering assortment of brutal treatments. Robinson casually states that in two or three cases he applied "a red hot wire" to a child's genitals. Joseph Howe advocated a treatment that involved a six-inch syringe ("Dr. Bumstead's syringes are the best") up the urethra.

The bitter irony here is that regularly spilling one's seed serves a valuable biological function. Sex physiologist Roy Levin explained to me that sperm which sit around the factory a week or more start to develop abnormalities: missing heads, extra heads, shriveled heads, tapered and bent heads. All of which render them less effective at head-banging their way into an egg. Levin speculates that that's why men masturbate so much: It's an evolutionary strategy. "If I keep tossing myself off, I get fresh sperm being made." Thereby upping the likelihood of impregnating someone and passing on your genes.

Though if conception is the goal, you don't want the sperm to be *too* fresh. Daily masturbation would deplete the number of sperm per ejaculate. Got to give the pinheads time to build up their ranks. To produce an ejaculate with optimum potential for fertilization, Levin recommends a holding time of five days.

france in the late-sixteenth and seventeenth centuries was really and truly a place where you did not want to be an impotent male. This was the era of the "impotence trial." Compared to the magistrates of these Reformation-era trials, Dr. Bumstead is the Gumdrop Fairy. When theologians elevated marriage to the status of a sacrament, impotence was likewise elevated, from a source of frustration to an actual crime. And because impotence was thought to arise from the intemperate spilling of one's seed, it was assumed that a man who could not get hard for his wife had been spending too much time doing so for others. Or, at the very least, that he was an immoderate masturbator.

Be that all as it was, the main reason a man's erectile capacity found its way into the courts was that impotence was a legal ground for divorce. Women seeking to escape a miserable marriage would accuse their husbands of it, with or without cause. If the wife won the case, the man would not only be fined and forbidden to remarry, but would have to return the dowry he had received from the woman's family.

For the husband to win his case, he had to prove himself capable of, as they say in modern-day erectile parlance, achieving and maintaining an erection. This meant a visit—often two or three or four visits—from a team of "experts" and examiners: as many as fifteen physicians, surgeons, and legal functionaries kitted out with their clipboards and pince-nez.

The defendant was examined in his home rather than in the courtroom, but it only moderately softened the humiliation. The team would arrive at the appointed hour and wait outside the bedroom until the defendant yelled through the door that he was ready for viewing. The examiners would file into the room and gather around the bed, whereupon the accused would pull back the bedclothes and show them what he had. These were tough critics. "We did find him in a state of erection upon our arrival," reads one report excerpted in Pierre Darmon's *Trial by Impotence*, "but he did not have sufficient attributes to consummate a marriage." How did they know? They leaned in and groped ("Touching this swelling, we felt it to be flabby").

Insult to injury, the examiners tended to wander afield from their appointed task, noting and commenting upon irrelevant anatomical quirks and afflictions. "We did perceive on the anus divers rather swollen haemorrhoids," snipes the report on one Jacques François Michel. Another defendant whose report is excerpted in Darmon's book, the Baron D'Argenton, was observed to have "no visible cullions [testicles], but as if a purse without sovereigns, . . . which did withdraw inside his person when he turned over, in such fashion that he had nothing left him but his member, and even this being far smaller than is customary among men. . . ."

Defendants occasionally resorted to extreme measures. The Marquis de Gesvres hired a theater troupe to perform an obscene vaudeville in his boudoir just prior to the arrival of the examiners. Others simply cheated. M. Michel, he of the swollen hemorrhoids, "uncovered himself only with the left hand while the fingers of the right pressed the root of the penis." (Too bad Robert C. Barrie hadn't been born yet. In 1907, Barrie received a patent for a hidden penis splint, which stretched the flagging organ along a thin rod between two metal rings, one concealed by "the gathered prepuce," the other by the man's pubic hair, "thus presenting a seemingly natural and unoffensive appearance to alley suspicion . . . [in] the opposite sex.")

A parallel absurdity took root around 1550. A medical expert of some repute made the claim that erection alone could not be considered sufficient proof of potency. Accused men would hereafter need to prove, in front of a panel of examiners, that they could mount their wife and ejaculate, as the medico put it, "into the appropriate orifice." Here it was the wife's genitalia that received the more rigorous scrutiny. "The woman is examined close up, to discover if she be more dilated than on the last inspection . . . (and if there be an emission, and where, and of what nature)."

Given that the wife stood to lose her case if her husband succeeded at his task, you had a situation that was equal parts rape and burlesque: "The husband complaining that his partner will not permit him to perform and does hinder intromission, his wife the while denying the charge and claiming that he would put his finger therein and dilate and open her by such means alone."

The year 1677, blessedly, saw the end of the era. A public prosecutor decreed the practice obscene and deplorable, and trial by congress was condemned to, as they say, the appropriate orifice.

this morning finds Dr. Hsu's patient subjected to a public inspection that, while not on the order of a French impotence trial, must be awkward to say the very least. The man reclines on a vinyl examining table, his arm behind his head. I am behind Dr. Hsu, trying to look as though I belong here. If not for the scatter of gray hairs, you would not guess the patient to be much over thirty. He looks at once bookish and athletic, the sort of man a mother will approve of. He is chatting with Dr. Hsu, who translates for me.

"He is feeling every morning tumescence. He hasn't had sex yet, but . . . No pressure, no pain." The patient is dressed in loose-fitting cotton shorts, which Dr. Hsu now instructs him to pull down. I pick up a journal on Dr. Hsu's desk and pretend to read it. It is published by the Taiwanese Association of Andrology, whose logo, I note, is a diverting variant of the snakes-and-sword caduceus: a penis flanked by free-floating testicles.

The man pulls down his pants without a flicker of embarrassment. Maybe he thinks I'm a nurse. Maybe it's different here in Taiwan. Maybe the population is more matter-of-fact about sex and nakedness than we are in the States. Taipei hotel rooms, I've noticed, have condoms the way American hotel rooms have shower caps and Bibles. Last night, channel surfing, I stumbled upon what appeared to be the local shopping channel. A man in a golf cap, looking bored, displayed a Nokia cell phone. Ordering information appeared on the screen. But rather than a narration about the excellent features of this telephone, the soundtrack was a song being sung by an agitated, single-minded woman: "I don't know your name, but it doesn't really matter. Let's have good hot long fast *wild horny dirty sex!*"

I find myself wondering about this patient. He seems too young to have problems with aging erectile tissue. What if his problem is psychological? Would therapy have been a better bet? Who knows. Perhaps he did not care to confront his emotional gremlins. Perhaps he preferred to blame his veins.

The desire to blame impotence on physiology rather than psyche is understandable. But caution is advised. You can't always trust science to get it right. Indeed, the first widespread surgical treatment of impotence was a farce of grand proportions.

7

The Testicle Pushers

If Two Are Good, Would Three Be Better?

When you are eighty-two years old and you have sixty-four wives, you need all the help you can summon. For Kamil Pasha, a vizier (bigwig) in the Ottoman Empire, help took the form of testicle consommé. The broth, which the Pasha quaffed daily, was made by steeping the danglers of robust young hoof stock. The vizier's enthusiasm for his vile bullion did not escape the notice of the harem obstetrician, Skevos Zervos. Zervos had been observing the Pasha's staff of eunuchs and musing on the feminizing effects of destroying a man's testicles. He began to hatch a theory of testes as the key to lifelong virility.

In between deliveries, Zervos began experimenting. He attempted to rejuvenate aging rabbits and dogs by grafting testicle tissue from younger specimens onto their own geriatric gonads. He went public with a paper in 1909, stirringly entitled "Curious Experiences with the Genital

Organs of the Male." The paper's aim was to popularize the technique so that eventually it would be available to all men.

The Pasha was not amused. He accused Zervos of plotting to restore his eunuchs' manhood. Fearing for his life, Zervos fled to Athens. There, in 1910, for the first and far from the last time, strips of ape testicle were implanted into a man. Zervos advertised the technique as a cure for both impotence and senility.

Word spread. By 1916, the nut graft had gone mainstream. In the first of two *JAMA* articles, G. Frank Lydston, a professor of genitourinary surgery at the Chicago College of Physicians and Surgeons, outlined the beneficial effects of tissue from a third testis—in this case, a human one—implanted in the scrotum beside the two that nature had bestowed.* Though an increase in "sexual power" and "vigorous and prolonged erections" were the most common type of claimed results, the secretions of the auxiliary gonad, in Lydston's view, erased many of the afflictions of advancing age: high blood pressure, senility, arteriosclerosis. At one point he described curing a twenty-two-year-old youth of, among other afflictions, the "frequent writing of incoherent, rambling dissertations on architecture." It seemed no ailment stood strong in the face of another man's testis.

---

*The scrotum enjoys a natural expandability that makes it a good candidate for use in skin grafts. Researchers at the Shriners Burn Institute in Galveston, Texas, have gone so far as to call the hairy sac "lifesaving."

There are images on the Internet of men with scrotums the size of those inflatable hop-along balls of my youth, but this strays well beyond normal expandability. These men have elephantiasis, and if you know what's good for you, you will not do a Google search of "scrotum" and "elephantiasis."

Lydston began by focusing on men whose own glands were obviously stunted or defective: the seventeen-year-old with no signs of virility and a testicle "about the size and shape of a small Lima bean," for instance, or the twenty-one-year-old saddled with "unsatisfactory and unpleasurable coitus" and testicles "the size of a navy bean." (Lydston had earlier compared a withered male gonad to a hazelnut, but wisely shifted to the bean family for his size comparisons—the word "nut" perhaps cutting a bit close to the Lydston bone.) As enthusiasm for the procedure built, Lydston began expanding his patient base, operating upon men with normal-sized testes as well. Early on, he had done the procedure on himself.

While Lydston had no trouble scaring up patients, finding donors was more problematic. Though a single testicle, like a lone kidney, is capable of taking on the duties of its absent teammate, rare is the man willing to part with a gonad for charity. Unable to secure the sex glands of virile young men, Lydston made do with the next best thing: the sex glands of *dead* virile young men. It appears he had a sympathetic friend at the city morgue, and that Chicago's young male accident victims around that time may have paid an additional price for their heedlessness. Most of Lydston's case studies described the unwitting donor as "dead of accident." Two hapless teens—one in 1918, one in 1915—succumbed to "a crushing injury" of the head, leading one to wonder if G. Frank Lydston occasionally prowled the back alleys of South Side Chicago armed with something larger and heavier than a navy bean.

Lydston insisted that dead men's testicle tissue would take, provided the operation was completed before the organ had begun to decompose; though as time wore on, he began to play looser with his rules. While the first donor testicle had been out of commission just six hours at the time of the operation, later organs sat on ice for thirty-six or even forty-eight hours.

This sort of delay wouldn't pass muster with French gonad grafter Serge Voronoff. The Russian-born émigré insisted that the gland must be transferred within seconds, lest changes in the cells compromise its vitality. Voronoff dreamed of a day when special hospitals would be set up, in which "candidates for glandular grafting will remain, in readiness to receive the required organs from the fatally injured cases who will be rushed thither."

The closest America came to such a facility was San Quentin State Prison's death row. In a series of experiments—about which we'll hear more shortly—San Quentin resident physician Leo L. Stanley plucked the gonads from some thirty freshly executed inmates and grafted them into thirty aged or "prematurely aged" inmates.

Meanwhile, Serge Voronoff, lacking contacts at prisons or morgues, was forced to come up with an alternate source of fresh testicle. Voronoff turned to apes and monkeys: chimpanzees, mostly, and the occasional baboon. The baboon, you may be interested to learn, carries a much bigger testicle than his TV-trainable cousin. So much so that Voronoff took to splitting the balls in two, parceling them out by halves to his patients.

The back half of Voronoff's 1925 book *Rejuvenation by Grafting* is a catalogue of case studies: page after page of aging men, pillars of society, who paid the charismatic Frenchman a prodigious sum to have slices of chimp testicle surgically installed in their scrotum. Along with his other medical innovations, Voronoff appears to have pioneered the medical Before and After shots. The Befores are invariably seated, but many of the Afters are action shots—seventy-year-old men in spats and cravats, striding and leaping across the lawn, demonstrating their newfound vigor.

By the patients' own testimonies, however, Voronoff's operation was something of a bust. "The genital coldness has changed very little," noted one. "I am somewhat more active; and the sexual function is very slightly stimulated. That is all," groused another. And another: "I am increasingly wretched."

Voronoff must have longed for the early, experimental days of his grafting career, when none of his patients complained or issued lukewarm reviews, because they were sheep. Beginning in 1913, he took the testicles out of more than one hundred young rams, sliced them, and grafted the strips into "miserable old beasts" who then became "young in their gait, bellicose and aggressive, . . . full of vitality and energy." In the Voronoff tradition, selected animals were posed for portraits, standing on seamless paper and looking, to my eye anyway, no different Before and After. Charmingly, Voronoff named his research animals, but not the way most people name animals. He named them the way people name perfumes, or in this case, something closer to an ill-conceived line of cologne: Old Ram No. 12, Old Ram No. 14, etc.

San Quentin's Leo Stanley had better luck than Lydston had with his human subjects. He published a paper in a 1922 issue of *Endocrinology*, summing up the effects of 1,000 testicular grafts. The donors were mixed: 20 executed San Quentin inmates and a party mix of common hoof stock, including goats, rams, boars, and deer.

The results were nothing short of astounding. Forty-nine of 58 asthmatics reported improvement, as did 3 of 4 diabetics, and 3 of 5 epileptics, 81 of 95 sufferers of "sex lassitude," and 12 of 19 impotent men. (These last two stand out as especially impressive—or something—given that he was dealing with an all-male prison population.) Even more astonishing, 32 of 41 men said they could see better, and 54 of 66 acne victims—apparently taking a daily blemish tally—reported a decrease "in the number of pimples." Dr. Stanley failed to mention something that David Hamilton, author of *The Monkey Gland Affair,* pointed out: that in exchange for participating in the study, the men were either paroled sooner or paid with cash. It's probable that they felt some pressure to tell the doctor what he wanted to hear.

Serge Voronoff kept data on his patients, too, but it was a perplexing mixture of bluster and failure. "In 26 to 55 percent," he wrote, "the physical and mental rehabilitation was accompanied by complete restoration of sexual activity." Well, was it 26 percent, or was it 55 percent?

It didn't matter. Testicle madness was in full bloom. Bars in the 1920s were serving a drink called the Monkey Gland, and shops in Paris sold ashtrays decorated with whimsical chimps with their hands on their genitals and the words "No, Voronoff, you won't get me!" By 1924, some 750 medical professionals and not-so-professionals were plying the gonad trade. These included not just the misguided and the deluded, but the intentionally deceitful. Prominent among the latter was John R. Brinkley, a Kansas-based surgeon with a diploma-mill M.D. and his own radio station. "A man is as old as his glands," went the Brinkley slogan, broadcast far and wide on KFKB. Every Sunday, a dozen or more somber men arrived at the Milford train station from points distant, to be processed for Brinkley's miracle four-part operation to restore sexual vigor with goat glands.

At one point, Brinkley was taking delivery on forty Toggenburgs a week, housed in a corral behind the hospital. He encouraged his patients to personally select their donor, like diners at a Chinese seafood restaurant being ushered to the aquarium. At a hearing described in *The Bizarre Careers of John R. Brinkley*, the enterprising quack was forced to admit that the operation was not only useless— the patient's immune system would attack the gonadal interloper, breaking the animal tissue down and eventually absorbing it—but often dangerous. Infections were common and patients were sometimes rendered sterile. Not to mention the ill effects on the donors, who were "fed to the coyotes" after surrendering their billyhood.

These days, the only animal testicles being implanted into men are silicone prosthetics called Neuticles— intended for neutered pets. Silicone is FDA-approved for breast implants and a host of other human body-part plumpers, but no one has paid the millions of dollars it would take to get testicular implants (for cancer patients) approved by the FDA—mainly because there aren't enough cases of testicular cancer to make it worthwhile. So men who have had a testicle surgically removed will sometimes order a Neuticle* and have a plastic surgeon install it. As long as the men don't mention that the prosthetic is for them, not their pet, the manufacturer can't get into trouble. These days Neuticles makes cosmetic gonads ranging from the Feline model to the Bull, accommodating, one hopes, most men's needs.

You may well be wondering why a neutered dog would need prosthetic testicles. A vet quoted on the Neuticles

---

*They can also, if they're the sort who likes to invite comment, order a Neuticles baseball cap or bathrobe, the latter suggesting a disconcerting scenario wherein the proud Neuticle owner suddenly throws open his robe for inspection.

The latest addition to the Neuticles promotional merchandise line is a Neuticles BBQ apron. It has been a sluggish seller.

Web site says the product "helps the pet's self esteem." I called Neuticles founder Gregg Miller to chat about the surprising notion of pet self-esteem. He talked about the day his bloodhound Buck was neutered. "I'll never forget it. He had just come home from the vet. He woke up. He went to clean himself, he looked down, and he looked back up at me. He *knew* they were missing. He was depressed for days." Miller concedes that Neuticles's healthy sales figures (157,000 pairs sold worldwide) may have more to do with male pet owners' hang-ups than with pets'—a fact supported by the not infrequent attempts to order Larges for, say, a beagle.

As far as I've been able to ascertain, the earliest genitalia pushers were the Chinese. (In this case, the gonad tissue wasn't grafted, but dried and made into pills or potions.) The *Chinese Materia Medica* of 1597, by the naturalist Li Shih-chen, promotes the penises of dogs, wild cats, and otters as treatments for impotence. Certain types of otters are deemed more effective than others: "In hunting for them women in spring go into the wilds in groups and the otter getting their scent jump upon them and cannot be removed except by throttling them to death. The penis of such an animal is considered very valuable." I had assumed that this was because the behavior denoted exceptional bravery and virility, or possibly because the throttling produced an impressive erection, but then I had to revise my theories, for Li concludes his entry with the following enigmatic fact: "The penis of an animal found dead clinging to a tree trunk is most costly." It's possible Li's grasp of otter biology was patchy. The freshwater otter, he writes, "is always male, and . . . it cohabitates with the gibbon."

Tiger penis as a cure for flagging libido is a more recent

addition to the Chinese pharmacopoeia. A 1993 report by the animal rights watchdog group Earthtrust describes a restaurant in Taichung, Taiwan, selling tiger penis soup to male diners for $320 a serving. (One penis makes soup for eight, the head waitress helpfully told a foreign television reporter.) However, there is no listing for tiger penis in the *Chinese Materia Medica*. Medicinal uses for the tiger's bones, flesh, fat, blood, stomach, testes (for scrofula), bile, eyeball, nose, teeth, claws, skin, whiskers, feces, and bones in the feces are listed, but none for the penis. Likewise, rhino horn* is listed as an antidote for typhoid, headache, carbuncles, "boils full of pus," and "the evil miasma of hill streams," but not impotence.

It is interesting to note that tiger penis is often taken in a potent wine, or brandy, which in and of itself might do the trick, should the man be suffering from simple performance anxiety. "Just prior to sex," states the Earthtrust memo, "the consumer takes a slug." It's clear I've been spending too much time with the *Materia Medica*, for upon first reading this, I pictured a man taking a garden slug as he might an aspirin. (Not entirely far-fetched: the sluglike sea

---

*Rhino poachers of yore were more resourceful than their modern-day counterparts, if no less evil. Li describes them "rigging up a rotten wood fence which the animals like to lean against." When the animals fall over "they cannot rise quickly and are easily killed." For sheer originality, though, nothing tops the tactics of hunters of the Moupin langur, an animal that grins "when it sees people" and "when it grins it draws its upper lip up over its eyes." Whereupon the hunter runs over and nails its lip to its forehead and smothers the creature. Don't worry, though; it's probably not true. Li also writes that the animal is ten feet tall and its feet are backward. The more you read Li, the more you wonder about his trustworthiness as a naturalist. One of his longest entries is Clear Liquid Feces of a Man, prepared by burying it in a jar underground to ferment. "The longer the time it is in the ground the better." Like, say, forever?

cucumber, like almost every other penis-shaped creature on the planet, is rumored to be an aphrodisiac.)*

Personally, I'd start with a prescription attributed to Aristotle. Seemingly under the sway of some ancient Gamebird Advisory Board, the author begins his list of potency restoratives with "pheasants, woodcocks, gnatsappers, thrushes, black-birds, young pigeons, sparrows, partridges, capons." The fowl eventually yield the stage to the chief "provokers" of erection: "watercresses, parsnips, artichokes, turnips, asparagus, candied ginger, acorns, scallions, shell fish," all to be used "for a considerable time." At some point, their efficacy probably became moot, as the men grew obese and sex began to take a back seat to the next shipment of gnatsappers.

d r. Hsu has a friend who owns a traditional Chinese medicine shop in Taipei, and he kindly agreed to take me on a field trip there. No gonads were for sale, and just one variety of penis: that of the deer. It looked like an antler, but Dr. Hsu was insistent. "Deer penis," he kept saying, as though dictating a letter. The proprietor would disappear into the back room and return with a stack of lidded cardboard boxes, like the shoe salesmen of my youth. Seahorses, fossilized oysters, antler chips, dinosaur teeth ("Just because they're hard," sniffed Dr. Hsu). The last remedy the proprietor brought down was a bundle of dried leaves

---

*But not if you step on one: A sea cucumber under assault expels the organs of its digestive tract through its anus (and then grows another set). Given that the sea cucumber's greatest threat is Asian cuisine, this "autoevisceration"—essentially gutting yourself and saving the chef the bother—probably needs to be rethought as a self-defense.

stacked like bills and tied with red cord. "Goat eat this leaf, mating one hundred times in a day," Dr. Hsu translated for me. The proprietor handed the sheaf to me.

"Thank you," I said. "But I don't really need any." I turned to Dr. Hsu. "Would you explain to him that I'm just doing research?"

A jolly-looking man browsing in the store walked over and joined our group. "You should ask for Emperor's Formula," he said. Dr. Hsu explained that an emperor during the Chin Dynasty, seeking to cure his impotence *or perhaps just doing research*, dispatched four hundred young men and women to an isolated island and instructed them to "try all the leaves and have fun!" The man and Dr. Hsu exchanged a few words in Chinese. The man looked at me, smiling.

"I'm just doing research."

dr. Hsu spent the drive back making sure I understood the limits of traditional Chinese medicines for erectile dysfunction. "Any formula," he said, "is not better than Viagra."

Viagra has an interesting history here in Taiwan. (And an interesting name: Ver Er Gang has been translated for me, variously, as "powerful and hard," "firm and strong like steel," and, my favorite, "steep and hot.") The week before the pill was to formally go on sale—as opposed to informally, on the black market, where it had been selling briskly for some time—women's rights activists took to the streets of Taipei, protesting its release. "Viagra has destroyed harmony at home and caused extra-marital affairs because wives cannot satisfy their husbands' sexual demands," one women's health advocate was quoted as saying in Malaysia's *Sunday Sun*. A gynecologist at Taiwan Adventist Hospital, where Dr. Hsu works, was also quoted, urging doctors not to prescribe the drug until they've been shown "written approval" from patients' wives.

I later learned that Viagra's release Stateside prompted a similar outcry from a faction of American women. Meika Loe, author of *The Rise of Viagra*, quotes five or six letters to Ann Landers. The sentiment runs more or less along the lines of No Name in Abilene, Kansas: "Please tell those smart-aleck scientists and those big drug companies to work on a cure for cancer and quit ruining the lives of millions of women who have earned a rest." Though Landers replied that these women represented a small percentage of her Viagra mail, the letters suggest that a little therapy tossed into the mix might not be a bad idea.

Of course, Pfizer Taiwan sailed ahead anyway. Its profits here and in mainland China have been, as they say, steep and hot. Perhaps the only locality where Viagra has proved a disappointment is the Wolong Nature Reserve in Sichuan Province. Wolong is home to part of China's dwindling panda population, as well as a group of captive pandas in a breeding research facility. Pandas have trouble reproducing in captivity. Some researchers describe it as a libido issue; some 60 percent of captive pandas show no interest in mating. Others seem to think that the males have erectile deficiencies, for in 2002, a middle-aged panda named Zhuang Zhuang was dosed with Viagra. "No result on him at all," the BBC quoted Wolong deputy director Wang Pengyan as saying.

Wolong researcher Guo Feng, also quoted in the BBC piece, took issue with his colleague. "You can't say Viagra has no results on pandas. That panda basically has no capability. In the last few years, we've given Zhuang Zhuang many chances but he simply can't do it."

It was unfair to single out Zhuang Zhuang, for male

pandas in general are bumbling lovers. "The male giant pandas do not know where to put it," a zoologist named Chen is quoted as saying in *Inside China Today*. "Sometimes they climb on the females' heads and start pushing." Seeking to enlighten clueless male pandas, Wolong staff set about making an instructional video, which the media gleefully dubbed "panda porn." The BBC even referred to the footage as "explicit," though given the animal's thick fur and diminutive penis*—erect, about as big as a man's thumb—it's hard to imagine that the pandas were able to glean much detail from the tapes. Likely more of a This End Up sort of deal. The staff tersely reported "improvement."

Pandas aren't the only ones for whom Viagra doesn't do the trick. About a fourth of the men who try it continue to have trouble getting hard. What's to be tried, short of a trip to Dr. Hsu's operating room? A few things. One can inject erection-producing drugs directly into one's corpora cavernosa, if one is the sort who takes kindly to sticking needles deep into one's penis. The injections prompt a speedy, impressive boner that typically lasts a couple of hours. Less typically it lasts a lot longer, transforming its owner from

---

*A fact that has spread well beyond the confines of panda reserves. The Urban Dictionary includes an entry for "panda penis" (meaning a small one) submitted by a contributor named Lew. Lew's usage example reads: "That girl said Matthew Reed had a panda penis." It is Urban Dictionary policy to reject definitions that include the full names of non-celebrities, leading one to assume that the panda slur was directed at some sports or entertainment figure named Matthew Reed, and not Matthew Reed the wedding photographer, or Matthew Reed the assistant professor with the research interest in the interaction of quinoid compounds with cellular macromolecules, or any of the thousands of other "Matthew Reeds" on Google.

Hercules to Priapus* and requiring an awkward visit to the emergency room. A newer, kinder, gentler way to deliver these drugs is via a urethral suppository known as MUSE (Medicated Urethral System for Erection).

And then there is the vacuum constriction device, a.k.a. the penis pump. This consists of a plastic tube, into which one puts one's flaccid penis, and a pump that pulls the air out of the tube. The resulting vacuum effect sucks blood into the organ, which is then trapped there by a constrictor ring. The penis pump was invented by a tire professional named Geddings Osbon, whose own frequent flats drove him to get creative with the company pneumatics.

As an indication of how thoroughly Viagra has deflated Osbon ErecAid company profits, witness the sad slogan "Make Your Second Choice Your Final Choice." One journal review from last year stated that penis pumps are suggested "mainly to elderly patients with occasional inter-course attempts."

Further along the fringes, there is acupuncture. While I was visiting, Dr. Hsu introduced me to a man who treats erectile disorders—though mainly premature ejaculation—with acupuncture. Hsu said 44 percent of men treated with acupuncture report improvement. This is less impressive

---

*I always assumed that Priapus was a god of something manly—war or shouting or chariot customizing—but in fact he was a god of fertility and gardens. One mythology Web site calls him the "the protector of all garden produce." Perhaps in defiance of the touchy-feely job title, he took to wearing robes slit high enough to display his enormous cucumber. Those caught robbing his garden were promptly sodomized. "If I do seize you . . . ," reads an epigram in Smithers and Burton's *Priapeia*, "you shall be so stretched that you will think your anus never had any wrinkles." Encyclopedia Mythica reports that outside of Rome, Priapus was "never very popular."

than it sounds, as 38 percent will report improvement if you give them a placebo treatment.

Acupuncture for impotence is more common on the agricultural scene. "Typically one treatment . . . with the moxa-needle technique is sufficient for impotence," writes the author of a paper on treating "overused" bull studs. Though it was not altogether clear whose impotence was being referenced—the bull or the man kicked in the balls by the bull as he tries to stick needles in its pizzle.

t he simplest, safest, cheapest treatment put forth for erectile dysfunction is to strengthen one's pelvic floor muscles. The late gynecologist Arnold Kegel was the first to discover the beneficial effects of pelvic floor muscle toning. Initially, Kegel's exercises were viewed strictly as a boon for women.* They helped reverse incontinence, and they helped women have more—and/or more intense—orgasms. Kegel promptly dubbed them Kegel exercises and went on the road to promote them, inextricably linking himself to pelvic clenching and ensuring an obituary that would gloss over all other salient aspects of his career.

In 2005, a group of professors at the University of the West of England reported in *BJU International* (*British*

---

*Kegeling has since been taken a step further, in the form of vaginal weightlifting. The idea being: You don't just flex your muscles if you want to build them up; you train with weights. I once tried the Feminine Personal Trainer for a story. It came with a slip of paper telling me not to be overwhelmed by its weight. I wasn't. I was overwhelmed by its size. Suffice to say, this is the only workout on Earth that calls for vaginal lubricant. The directions tell you to insert and contract, causing the FPT to rise up inside you until all that can be seen protruding is a doorknob-shaped piece of steel, as though you are giving birth to a hardware store. I use mine as a paperweight.

Journal of Urology) that Kegel exercises could cure erectile dysfunction. After three months of twice-daily Kegeling sessions, a group of impotent men showed significant improvement as compared with a control group. An "independent blind assessment"—the logistics of which I cannot begin to contemplate—determined that 40 percent of the impotent men, compared with controls, had regained normal erectile function. In a further 35 percent, the condition had improved. The paper points out that Kegel exercises were more effective in the younger study subjects, subjects whose erectile dysfunction was less likely to stem from the usual band of brigands: diabetes, atherosclerosis, fibrosis.

Impressed and curious, I sent away for the instructional video referenced in the paper, which was produced by Grace Dorey, one of the paper's authors, and a physiotherapist named Kevin Foreman. The video begins with an introduction by Dr. Foreman (who has a Ph.D., not an M.D.). He wears a peach-colored tie and sits against a backdrop of a plain curtain. Neither his chair nor the table he sits at can be seen. Not even a lone potted plant has been recruited to share the frame with Dr. Foreman and his tie. The colors are garishly distorted, so that Dr. Foreman is the pink of an Easter ham.

Dr. Foreman explains that one pair of pelvic floor muscles runs along the sides of the penis like guy wires and helps hold it erect. Another pelvic floor muscle encircles the penis and, when tightened, puts pressure on the dorsal vein and helps trap blood inside the penis. Toning these muscles should, in theory, help men have firmer, longer-lasting erections.

The how-to portion of the video starts innocently enough, with anatomical drawings on a sketch pad. Viewers are instructed to simultaneously squeeze as if pretending to stop the flow of urine and "tighten the back passage

as if to prevent wind escaping." I recognize this particular euphemistic phrasing from one of Ms. Dorey's patient
instruction booklets (a list of which she had emailed me):
*Prevent It! A Guide for Men and Women with Leakage from the
Back Passage.*

Abruptly, the scene shifts to a man's naked hips on a
rumpled white sheet. He lies on his back, knees up, thighs
spread, back passage clearly visible. Understandably, the
man's face is out of the frame. A catalogue description of
the video states that the exercises are demonstrated by "a
male model," but I suspect it's Foreman. A budget so small
that it cannot accommodate a potted plant does not easily accommodate modeling fees. Though I could easily be
wrong. I could be talking out of my back passage.

A voiceover puts the pelvic floor muscles through their
paces. At one point, we hear "Tighten! Tighten! Tighten!"
in a ringing exclamatory voice. My husband, in the kitchen,
heard this as "Titan! Titan! Titan!" and comes in to check
out what he assumed was some deliriously campy, low-
budget sci-fi pic.

"Whoah," says Ed upon encountering Dr. Foreman's
package.

"Notice the lifting of the scrotum," the narration continues. Next the model demonstrates the exercises from a
standing position, and here the camera zooms in dramati-
cally. Ed and I lean back into the sofa cushions.

"Tighten! Tighten! Tighten!"

Part II was entitled "Post-Micturition Dribble." "This
is a dribble of urine that occurs after you've finished going
to the toilet," explains Dr. Foreman. The video cuts to a
shot of a boxer-clad bottom backing away from a toilet.
The skin tone situation has worsened, such that the man
appears to be wearing a hot pink leotard under his boxers.
The camera displays a close-up of a dampened blotch on

the fly of the boxers. "It can be as much as an eggcup-ful and can be very embarrassing." Is this an estimate, I wonder, or did Dr. Foreman recruit someone to dribble into an eggcup?

Given Foreman's findings, why aren't Kegel exercises the toast of the urological community? "That's one study," says American Urological Association spokesman Ira Sharlip when I ask him about it. Indeed, a search of the medical journal database PubMed unearthed no attempts to replicate the West England Kegeling study. Which seems like a shame. (A general Internet search did turn up a few self-help sites that mention Kegeling as a way to control premature ejaculation, something to which Dr. Foreman also alluded.)

Kegeling seems to me to belong in the same box with oat bran and prayer and vitamin pills, the big box that says CAN'T HURT, MIGHT HELP on the side of it. Certainly, it would be worth trying before moving on to the final frontier, the treatment of last resort: the artificial erection.

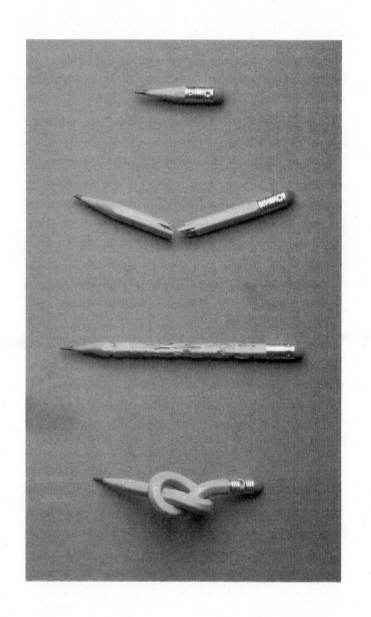

8

Re-Member Me

Transplants, Implants, and Other Penises of Last Resort

he AMS Malleable 650 Penile Prosthesis is a high-profit item with a steady demand. You could do worse, in life, than to be an AMS sales rep. On the downside, your sales patter would need to include the phrase "better concealment with less springback." You would have to listen to yourself saying "The enhanced rigidity reduces the possibility of buckling during intercourse." There would be days, perhaps even most days, when you would find yourself in a group of strangers, holding a silicone penile implant up to the fly of your chinos.

For the AMS sales rep visiting Dr. Hsu's office at the Microsurgical Potency Reconstruction and Research Center, today is one of those days. Worse, it's a day when the strangers are myself and Dr. Hsu's wiseacre colleague Alice Wen. Alice is serving as my interpreter. The rep is demonstrating how you get erect with a set of 650s—one in each erectile chamber—inside your penis. It happens faster than the real deal. You just bend it into position like a gooseneck lamp.

There are also inflatable, rather than malleable, models. Here you don't bend the penis, you pump it up. The surgeon implants a small bladder of saline* (or air) above the pubis bone. This gets pumped into the implant by means of a hollow, squeezable bulb implanted in the scrotum and attached to the prosthesis by a plastic tube. Inflatables are more popular because—unlike a malleable implant—they enlarge the girth of the penis, as would happen in an unaided erection. To many men, it seems more natural—except, of course, for the scrotum-squeezing aspect of the event.

"So who does the pumping?" Alice is making what psychologists call the distress face. "Anyone" is the answer. Whoever wants to. The guy, his date. Very occasionally, a visiting stranger. Sex researcher Cindy Meston tells a story about the time Irwin Goldstein, then at Boston University's Center for Sexual Medicine, made her pump up one of his post-ops. "I was in Boston for a conference. I had the flu, I was throwing up all morning. Irwin was all excited about this new pump he'd installed: 'You have to see this, Cindy!' He drags me over to his office, and there's this enormous man with no pants on. Irwin's going, 'Go on, Cindy, pump it up!' And I'm going, 'Oh, no, Irwin, *please*, not today. . . .'"

Who would have this done? "It's fairly popular," says Goldstein, a urologist who is the editor of the *Journal of Sexual Medicine*. (The global total, to date, for AMS implantations is 250,000.) "But it's a third-line therapy." In other

---

*To those who would say this is unnatural, I direct you to consider the male shark, whose sexual apparati are also erected with saltwater. Shark "claspers" fill with seawater before the predator mates.

words, doctors try less radical treatments first. Implant surgery is intended for men whose erectile tissue is irreparably damaged and fibrous. Because if it wasn't that way before the operation, it will be afterward. The prosthesis basically reams the erectile tissue on the way in.

Despite the reaming, an implant recipient can have orgasms and ejaculate. "The rigidity function—which is now being borne by the implants—has nothing to do with desire and orgasm," explains Goldstein. Erection, orgasm, and ejaculation are independent events. A man can have an orgasm—or even multiple orgasms*—without ejaculation, and he can have an orgasm and/or ejaculation without an erection. An implant only affects erection. Goldstein: "If you can play the piano before the implant, you can play the piano after the implant."

Most prosthesis patients are older men. In about twenty minutes, Dr. Hsu will be inserting the AMS 650 into a seventy-year-old man whom I will call Mr. Wang. The reason for the operation sits in the waiting room: the new wife. Mrs. Wang is forty.

t he first time an implant—basically a strip of cartilage—was installed in the penis of an impotent man was 1952.

---

*In 1989, a team of psychiatrists at SUNY Downstate Medical Center interviewed twenty-one men who did so regularly. Some had always been able, and some had come upon the ability later in life. The latter group combined those who had taught themselves, using a variety of techniques generously covered on the Internet, and those who stumbled upon it accidently, such as the fifty-nine-year-old skeet shooter who had continued thrusting on behalf of his wife and surprised himself with a second orgasm. Whereupon he exclaimed, "Doublee!"—described by the paper's authors as a term for shooting two clay pigeons with a double-barreled gun, an event even rarer than male multiple orgasm.

The patient is described—this being the *Journal of the South Carolina Medical Association*—as "a 23-year-old Negro veteran of World War II." Ironically, the young man became impotent as a result of being the opposite of impotent. Three months before, he had shown up at the veterans hospital with an erection that had refused to go down for two days and two nights. The doctors surgically drained the corpora cavernosa, and the operation resulted in a constricting scar in his erectile tissue, such that it was no longer living up to its name. Insult to injury, when he returned to complain about his impotence, the doctors refused to take him at his word. They had him masturbate in front of them. When this "failed to cause any visible or palpable erection of the penis," Dr. Buford S. Chappell signed him up for the world's first penile implant. Chappell does not specify who—or what—the cartilage came from. Nor does he mention whether the patient knew he was a guinea pig.

You got the sense that Chappell's summation—"He has ejaculations although intercourse is not as pleasant as before"—wore a heavy sugar-coat. Chappell included an After drawing of the penis, which "now hung in a semierect position that allowed the comfortable wearing of clothing." Based on the drawing, it was difficult to imagine wearing anything other than a caftan comfortably, and then only if you were comfortable with constantly appearing to have an erection, something that if I were a young African-American man in South Carolina in 1952, I might not be.

alice and I are joining Dr. Hsu in the operating room today. Mr. Wang is resting with his eyes closed. The operation is being done with local anesthesia, backed up by acupuncture. The penis is in good hands: Dr. Hsu has done more than a hundred prosthetic implantations. "Exactly one hundred eighteen," he says, vigorously scrubbing his hands and forearms at the sink in the corner of the operating room. "Only two extrusions." Alice raises her eyebrows above her surgical mask. "You mean it . . ."

Dr. Hsu holds his hands up in front of a nurse for gloving, a word I can no longer, since chapter 6, type happily. He nods. Meaning it pokes through the end of the penis. This tends to happen during vigorous sex in certain positions. "That's why we tell patients: No woman on top."

"Ohhh, no." This from Alice.

Dr. Hsu makes a short incision where the pubic hair would be if the patient hadn't been shaved. (This way the incision scar is hidden; implants can also be inserted through the tip of the penis.) He picks up a steel rod of approximately the same dimensions as the implant. This rod, called a Hegar's dilator,* will be used to stretch each corpus cavernosum to ready it for the implant. It slides in fairly easily, though not as easily as I would have preferred, catching here and there and requiring a firm push.

Dr. Hsu's nurse is unwrapping the second implant, for the other erectile chamber. This one does not go gently. The insertion is done in two stages. One end is submerged down to the pubis bone. That leaves several inches of implant sticking out of the incision like a flagpole upon conquered lands. Dr. Hsu hairpins the protruding part

---

*Medical dilators have been around since the early 1800s, mostly under the name *bougie*. There is a bougie for widening most every tube in the body. There are cervical bougies, urethral bougies, esophageal bougies, vaginal bougies. There is even, yes, a sinus bougie. Not only is there an anal bougie, there is a neoanus bougie, used for dilating a new asshole. Hegar refers to the man who invented this particular dilator. More specifically it refers to Ernst Ludwig Alfred Hegar, and if ever a name called out for anal dilation, it's that one.

in half and then tries to feed the remaining end into the incision and push it toward the organ's tip,* straightening it as it enters. It seems to be stuck. There's a kinked inch of implant protruding from the fleshy incision. Dr. Hsu presses on the kinked rod. The novelist Martin Amis once described an impotent character's attempts at intercourse as being like trying to feed an oyster into a parking meter. This is like trying to put a parking meter into an oyster.

Dr. Hsu pulls out the 650 and starts anew. Alice has stopped watching. Mr. Wang, incredibly, is napping. This time it goes in with minimal wrangle, and Dr. Hsu sews up the incision. It looks like a penis again, but longer and fatter than it was an hour ago. Dr. Hsu does a dry run, making sure the implant bends properly and holds its erect position. He bends it up into the familiar silhouette, and then lets go of it. It stays as he left it, a cooperative flesh Gumby limb. Then he pushes it down out of the way, like those exercise gizmos that can be stowed flat under the bed. Mr. Wang won't *become* erect, he'll just suddenly *be* erect. His hydraulics have been swapped for an ON/OFF toggle.

What Mr. Wang has sacrificed today is his organ's natural retractability. The adjective *flaccid* will never again apply. In its place are the adjectives *bulky* and *conspicuous*. Mr. Wang will appear to be going through life at half-mast.

---

*A word of apology to the male reader. There is no way for me to adequately appreciate how uncomfortable this may be making you. As a consolation, I promise never to witness and describe the insertion of a Disposable Internally Applied Penile Erector. (Yet one more way to stiffen an unstiff penis.) U.S. Patent 4,869,241 describes a stiff, hollow (so semen can come out) plastic tube designed to be "slidably placed" in the urethra. By the urethra's owner. Writes patent holder John Friedmann—clearly of tougher stuff than my husband, Ed, who actually crossed his legs as I read this to him—"One merely slides the support tubing down the urethra."

Too bad he doesn't have a pair of underwear that exerts significant inward retentive pressure. I am borrowing the wording of the team of inventors listed on the patent for Men's Underwear with Penile Envelope.* The patent nowhere states that either of the inventors—who share a last name—had a semirigid penile implant that was causing embarrassing trouser bulge. Nor does it state that the other inventor exerted significant pressure to do something about it. I am, as they say, thinking outside the penile envelope. Just guessing.

Mr. Wang's penis resembles a normal erection, but I find myself wondering if it feels that way. Does it feel like a blood-engorged penis, or does it feel like a penis with two silicone rods in it?

"May I squeeze it?"

Alice looks at Dr. Hsu. Dr. Hsu looks at the patient, whose head is hidden behind a green curtain of surgical sheeting. He is awake now, but he speaks no English and his crotch is still numb. He'll never know. Dr. Hsu steps away from the operating table and pulls a pair of latex gloves from a box on a counter behind him.

"Mary, you have traveled a long way. You can do whatever you want."

It does not feel entirely penislike, but at the same time, it does not feel inferior. It's sort of bionic-seeming.

---

*Close to but not quite the world's most embarrassing underthing. First prize must go to the Deodorizing and Sound-Muffling Anal Pad. The patent's background material details the sad decline of the human anal sphincter muscle, whose gripping capacity fades as we age. The absorbing layer is said to "trap the sound of a flatus," as though one might later drive it to a less populated area and release it.

The Anal Pad should not be confused with a prior invention called the Anal Napkin, which, in turn, should not be confused with the dinner napkin.

Though the exterior lacks the steely* feel of a true erection, the interior is hard, harder even than a natural erection. And so it stays until everyone is finished. I can understand why, for someone who has exhausted all other possibilities, implants could be a welcome relief. So are they? And what do the partners have to say?

In one study, 76 percent of the men were satisfied with the rigidity of their new, malleable organ. Another survey, of 350 men with inflatable implants, showed a satisfaction rate of 69 percent. In another, similarly sized study, 83 percent were satisfied—but only 70 percent of their partners were. ("Because they want to go on top," surmises Alice Wen.) The most common complaints were that the implant caused pain or that it looked unnatural. Women occasionally complained that the head of the man's penis was cold (a condition known as "cold glans syndrome").

Obviously, satisfaction rates vary depending on the type and brand of implant. Less obviously, satisfaction rates vary depending on which one of a man's wives is weighing in. In a study entitled "Satisfaction with the Malleable

---

*Not entirely an exaggeration. The collagen fibers surrounding the corpus cavernosum of an erect penis are as stiff, by weight, as steel. I learned this in 1999, while interviewing Diane Kelly, then at Cornell University, the planet's lone expert on the biomechanics of the mammalian penis. The fibers are arranged in two layers, one perpendicular to the other, which keeps erections from bending or ballooning out of shape when they're squeezed. If you use enough force, however, a penis will buckle. "Penile fracture" is the preferred term. It refers to a ruptured corpus cavernosum rather than a broken bone. Humans don't have penis bones. Dogs do, and chipmunks and muskrats and various other mammals, all of them represented in the fabulous Smithsonian Institution penis bone collection that languishes, tragically, in an off-site storage facility. The largest penis bone is that of the walrus. The Inuit call it an oosik and used it as a war club.

Penile Prosthesis Among Couples from the Middle East," some of the men—who hailed from Libya, Egypt, Sudan, Yemen, Algeria, and Saudi Arabia—had either two or three wives. Table 4, a journal table quite unlike any other, lists the men's and women's assessments in separate columns, and is split crosswise into sections for "3-Wives Polygamy" and "2-Wives Polygamy." While in six of the nine polygamous couples, the men basically agreed with their wives, three couples' assessments bespoke stormy days in the marital tent.

Polygamy No. 4, for instance. The man reported being "satisfied," as did one of his wives. The other two wives were, respectively, "dissatisfied" and "very dissatisfied" with the implant, "to the point of strong desire to remove it." I pictured wives two and three whispering conspiratorially, kitchen knives hidden beneath their abayas. You may add to the general climate of marital discord the fact that 64 percent of the men had kept the operation a secret from their wives—possibly because 94 percent of the men hadn't been told by their surgeon that their erections would be unnatural. The capper: A couple of years after the men's surgeries, Viagra became available in most of these countries.

The careless reader might be tempted to draw a conclusion from the preceding paragraphs, and that is that polygamy causes erectile dysfunction. *Au contraire!* In 2005, anthropologist Ben Campbell traveled from Boston University to the far fringes of Kenya to chat with Ariaal tribesmen about their erectile function. One of the things he discovered was that men with multiple wives had lower rates of age-related erectile decline. Of course, this is not to say that an extra wife will prevent you from developing ED. It is far more plausible that a man whose penis is in working order is more likely to take on the sexual freight of multiple wives. Compared, that is, to an impotent man. Who would have to be mad, or maybe Libyan.

rather than shore up a broken penis, might it be possible to simply install a new one? If a hand or face can be transplanted, why not a penis? Surgeons have, in fact, considered it. Danish surgeon Bjoern Volkmer said in an email that the topic came up some months ago with regard to a young patient whose penis had been partially amputated to remove a malignancy. One problem, Volkmer said, is that erectile tissue can react to trauma by growing the same sort of tough, nonelastic, fibrous tissue that contributes to impotence. This includes the trauma of attack by one's own immune system, an inevitable side effect of transplanting someone else's tissue into or onto your body. Immunosuppressive drugs would mitigate, but apparently not prevent, the problem. (The other, larger hurdle, with cancer amputations, is that the immunosuppressive drugs needed to protect the new penis would leave the patient defenseless against the cancer.)

But if you happen to be impotent because someone has *cut off* your penis, then the microsurgeons can help you. The world's most experienced penis reattachment surgeons can be found in Thailand, where, during the 1970s, an estimated one hundred vengeful Thai wives, spurred by media coverage of a prominent 1973 case, sliced off the penises of their adulterous husbands as they slept. When a suitably equipped microsurgeon was on hand to reattach the errant appendage, the men were able to resume philandering within a matter of months. Though probably with reduced success: The penises, though operative, were shorter, numb, and often only partway erectable.

The most serious complication, in the Thai attacks, was

infection. Two of the wives flushed the penises down the toilet, forcing their husbands to grope for their lost manhood inside the septic tank. (Incredibly, both were found, cleaned, cleaned some more, and reattached.) More commonly, the women would hurl the penis out the window. In the cases described in "Surgical Management of an Epidemic of Penile Amputations in Siam," all the recovered penises were "grossly contaminated."

Better that than eaten by livestock. Many rural Thai homes are elevated on pilings, with the family's pigs, chickens, and ducks tending to mill about seeking shade in the space underneath. It is not, oddly, the pigs, but rather the ducks, that the castrated Thai must worry about. The paper does not provide the exact number of penises eaten by ducks, but the author says there have been enough over the years to prompt the coining of a popular saying: "I better get home or the ducks will have something to eat."

And then there are the castrations wherein the blade and the stalk belong to the same man. One of the Thai case reports was that of a husband whose wife had complained about his failings in bed, whereupon he walked into the bathroom and severed his penis with a straight-edge razor. (While the Thai women in the article almost without exception used kitchen knives, the autocastrating male tends to reach for his razor. Or, in the case of one Thai farmer, a *shovel*.) The remorseful wife rushed both husband and penis to the emergency room—the latter wrapped, like an exotic lunch, inside a banana leaf.

A small but unsettling subset of autocastrations are the product of religious delusion. The New Testament contains a troublesome passage about celibacy (Matthew 19:12). In the passage, Jesus is ticking off all the kinds of eunuchs in the world. "There are eunuchs born that way from their mother's womb, there are eunuchs made so by men, and there are eunuchs who have made themselves that way for the sake of the Kingdom of Heaven." In 1985, a thirty-one-year-old Australian man added himself to the last category. It was an especially tragic case, as this man was already more celibate than many priests. He had never had sex, never had a girlfriend (or boyfriend). He lived with his parents, where, quoting the case report, "every spare moment was spent sitting in his room reading the Bible from cover to cover."

Except for the occasional five or ten moments that he devoted, with tremendous guilt, to masturbating. Worried that the keys to the Kingdom would be withdrawn, or the locks changed or however that works, he decided to atone. Employing the skills he had picked up while castrating bulls on his father's farm as a boy, he opened his scrotum with a razor, cut out his testicles, and flushed them down the toilet. The author of the case report interprets the disposal as a sign of resolve; ambivalent autocastrators often "bring their organs with them to the hospital." Indeed, this man made no effort to retrieve his testes and voiced no regret for the act. He also refused testosterone replacement therapy, and has no doubt made great strides in the church choir.

I will leave you with the story of a fifty-six-year-old government officer in India who told emergency room doctors that he had cut off his penis in order to cure a long-standing case of incontinence. One of the physicians, who wrote up the case for the *Indian Journal of Psychiatry*, suspected deeper mental turmoil was at play, as the patient had earlier jumped into a well, in what was ruled an attempted suicide. Though given this man's flair for therapeutic overkill, it's possible he was merely thirsty.

A penis amputation is not a cure for incontinence but it was, for one perplexed seventy-year-old man, a cure for his impotence. The newly invigorated member was a phantom.

Phantom limbs are a common consequence of arm and leg amputations; owing to peculiarities of the nervous system, the sensations that existed in the limb beforehand often persist after the surgery. Occasionally, this happens with mastectomies—the phantom breast even seeming to "swell" at certain times of the month—and with penectomies (seven of twelve cases, according to one survey). Phantom expert V. S. Ramachandran, of the Brain and Perceptual Process Laboratory at the University of California, San Diego, once had a patient with a phantom appendix. So painful was it that he had trouble believing his surgeon had removed the real one.

In the case of the seventy-year-old, the phantom erection was so vivid that the man would bend over and "check for its presence." It must have been a bittersweet victory: to feel erections after two years of impotence, yet have no penis with which to take advantage of them. It's a urological rendering of the O. Henry story about the woman who cuts off her long hair and sells it to buy her husband a watch fob for Christmas—not knowing that he has pawned his watch to buy a set of combs for her hair.

The phantom erections eventually stopped when the penis amputee, at seventy-four, was shot in the back and paralyzed from the waist down.

9

The Lady's Boner

Is the Clitoris a Tiny Penis?

a woman in an MRI tube has few secrets. The man at the control console knows the size of her heart and the contents of her womb. He knows if she's had her breasts enlarged or her stomach stapled. He can see into her bladder and knows whether she's wishing she'd stopped by the restroom before climbing onto the exam table.

Ken Maravilla, a University of Washington radiology professor, knows all these things about Meg Cole★ (bladder contents: half a cup). Very soon, he will also know how stimulating she finds the X-rated video that he has arranged for her to watch while she gets an MRI. Before you too arrange to have your next MRI done at UW, you should know that not everyone gets the Maravilla treatment. Only study subjects.

★Not her real name.

Maravilla has a side interest in sex research. His work has shown that MRI can provide an unambiguous measure of how much blood is in the tissues of a woman's clitoris. As it does with men, sexual excitement ushers more blood to a woman's genitals. Clitoral blood volume, then, should yield a simple, dramatic portrait of what Masters and Johnson oh-so-appealingly called "mounting readiness." Maravilla has found that, on average, women's clitorises hold twice as much blood while they are watching porn than when they are watching, say, footage of a Space Shuttle launch.

"Mary, we don't use the term 'porn,'" Maravilla says quietly. "We say 'erotic videos.'" (Or, when we're feeling especially defensive, "VES," for visual erotic stimulation.) Maravilla, sixty, is a slim, gracious man. His hair is cleanly cut and contoured. He speaks easily about sex, but is sensitive about the porn thing. This is understandable: His proposal was originally turned down by the university's human subjects review board. Maravilla had to sit down with them, convince them that "it was above-board, that there was nothing voyeuristic going on."

Of the many ways to quantify a woman's sexual fires, MRI is the least intrusive, in that nothing need be inserted, suction-cupped, or otherwise affixed. This is sexological measurement at its most demure. Cole lies on her back with a radio frequency coil laid lightly, like a heating pad, over her hips. She is given a pillow and a blanket, and the lights are turned down. It's like first class on a British Airways flight to Europe: downright comfortable under the circumstances.

Cole has what the *Diagnostic and Statistical Manual of Mental Disorders* calls female sexual arousal disorder (FSAD): She is regularly in the mood for sex, but her body doesn't respond to the preliminaries. To be more precise, it doesn't respond the way she'd like it to. If it wasn't a problem in Cole's eyes, it wouldn't be a problem in the eyes of the *DSM*. Part of the diagnosis is that the condition causes "marked distress or interpersonal difficulty."

FSAD is the ladies' edition of ED (erectile dysfunction). It is distinct from FOD (female orgasmic disorder) and HSDD (hypoactive sexual desire disorder, or low libido).* Confusingly, there is also female sexual dysfunction, or FSD, but this is simply the catch-all term for anyone who has one—or a combination—of these conditions. Lack of desire (HSDD) is the most common of women's sexual complaints, and we will get to this later. (Monkeys will be involved. Do not change the channel.)

Female sexual arousal disorder is the least common. Women with FSAD and nothing but FSAD are difficult to find. The coordinator of tonight's study, Joanna Haug, had to interview 140 women to find the volunteers who comprise the study's subject group. Unexpectedly, Haug finds many of her subjects by running ads in the sports section of the *Seattle Times*. "I used to run them in the food section," she says, "but I kept hearing women tell me they heard about the study from their husband." Men are often more troubled by their wife's sexual responses, or lack thereof, than are the women themselves.

Cole is not a stereotypical "nonresponder." She is

---

*And from HAFD, hyperactive acronym formation disorder. The condition has reached epidemic proportions in the sex research community. Cindy Meston staged a quiet parody in her days as a postdoc at the University of Washington. She had the task of composing a questionnaire to screen patients to see if they were promising candidates for surgical correction of a crooked penis (due to Peyronie's disease). The surgery repairs the crook but takes as much as an inch off the length. Meston called the questionnaire the Washington Examination of Expected Negative Identity Post-Peyronie's: the WEENI PP.

not conservative or inhibited. She is the opposite of these things. When I came in, she was chatting with Haug about the best Seattle sex shops. A friend of hers teaches a bondage safety class sponsored by a local police department. "People kept getting hurt," she is saying. "I guess it was monopolizing too much of the officers' time." If a woman like Cole is having trouble getting aroused, it seems reasonable to look for a physiological explanation.

Haug sits beside Maravilla in the MRI control room, which is actually labeled "Control Room," just like in a James Bond movie. She is a likable, no-nonsense English-woman in pinstripe pants and a ponytail. At Maravilla's signal, she presses a button to activate a conveyor that will move Cole into the MRI machine. The conveyor carries her toward the magnet feet-first and cinematically slowly.* James Bond inching toward the spinning saw blade.

Haug selected the sexually explicit clips for the study. I ask her where she got them, thinking, I don't know, that there's a supply catalog of explicit video clips produced specially for sex researchers. Haug blinks at me. "Sex shop." Joanna Haug has more interesting business receipts than the rest of us.

Maravilla centers an image of Cole's clitoris on the screen. The image bears no resemblance to the vague nub one generally pictures. This is because we're looking at the whole organ, not just the one-tenth of it that is visible to the eye. Maravilla takes me on an underground tour. "Here

---

*Other things enter MRI tubes less slowly. Anything magnetic is subject to the projectile effect and, if brought too close, will abruptly lift off the ground and hurtle through the air toward the giant magnet at up to 40 mph. This has, in the past, included ladders, floor buffers, laundry carts, IV poles, and, in July 2001, an oxygen canister that fatally struck a six-year-old boy as he lay in the tube.

are the crura," he says, pointing to a pair of matched arms that branch away from the tip like halves of a wishbone. "And this is the glans," he says, as though Cole has a penis. In a sense, she has. Time out for a short primer.

I can recall, many years ago, being told that a clitoris* is a vestigial penis. The feminist in me, who is small and sleeps a lot but can be scrappy when provoked, took umbrage at this description. I resented the implication that men have the real deal, while women make do with a sort of miniaturized, wannabe rendition.

But it is true. Male and female fetuses both begin life with something closer to a clitoris. The male's expands into a penis, while the female's remains more or less as is.

Even in their adult forms, the two organs have much in common. The clitoris, like the penis, ends in a sensitive, nerve-dense, pleasure-yielding bulb of tissue called a glans. Like the penis, the clitoris has a shaft, and that shaft contains a pair of expandable chambers called corpora cavernosa. It also has a prepuce, or foreskin, just like the penis does, and if you draw it back you may, just as with the penis, discover a wee cache of smegma. Robert Latou Dickinson, a pioneer in the field of female smegma, described it in his *Atlas of Human Sex Anatomy* as "tiny, hard pellets, white and

---

*Let's get used to this word, because there are few likable substitutes. In a study of male and female genital slang carried out at five British universities, respondents came up with 351 ways to say penis (e.g., veiny bang stick, custard chucker, one-eyed milkman, bishop) and only three for clitoris: bean, button, and the little man in the boat. The authors felt this reflected society's disregard of female pleasure, which is probably true, but I simply bemoan the lack of useful synonyms. The third one is well-nigh unusable, as anyone from this side of the Atlantic would assume the reference was to the Ty-D-Bol Man.

glistening." The text includes a thumbnail illustration of three such pellets, placed alongside a clitoris, for scale. The drawing is dated 1928 and labeled, in Dickinson's neat, boyish calligraphy, "Smegma."

And yes, a clitoris expands when its owner is aroused—though not as quickly or extravagantly as does a penis. Masters and Johnson filmed dozens upon dozens of clitoral erections: Responses to vibrators, to fingertips, to "stimulative literature," to intercourse. (Filming the clitoris during a missionary position coupling was of course problematic, as the male missionary is in the way. In this case, the artificial coition machine was called in to pinch-hit.)

Masters and Johnson focused on the glans—the touchy little bean that is the visible portion of the clitoris. The pair found that although the glans occasionally gets much bigger, expanding to as much as twice its normal size, fewer than half the clitorises enlarged to a degree that could be detected by the unaided eye.

MRI tells a different story. Maravilla's work has shown that when you factor in the blood volume of its hidden portions, the aroused clitoris routinely doubles in size.*

---

*A clitoris does not, of course, ejaculate semen. But some women—40 percent according to a 1990 survey of 1,292 women—do, like men, expel a substance from their urethra during orgasm, especially orgasms from stimulating the G-spot region. The nature of this "ejaculate" has been the source of extended debate—*urine or not urine?*—and diverse scientific inquiry. One woman devised a home experiment in which she swallowed a tablet that dyes urine bright blue. She then "inspected her wet spots," which she reports were either colorless or faint blue. One research team collected specimens of "the expulsion" and asked outsiders to characterize it. It is a testimony to the generosity of the human spirit that these volunteers both smelled and tasted the specimens. (Several specified that it "had no urine taste" without further specifying how they would recognize this taste. One likened it to watered-down fat-free milk.)

Given the anatomical parallels between penises and clitorises, it might seem reasonable to think of women's arousal problems, like men's, as a blood-flow issue. Indeed, if you look in the sexual medicine journals, you will find papers on clitoral priapism, nocturnal clitoral erections,* and, yes, "clitoral erectile insufficiency." The team of researchers who coined this term are, unsurprisingly, urologists. They theorized that when older women have trouble getting aroused, the culprit might be—as it sometimes is in older men—clogged arteries. To test their theory, they induced atherosclerosis in a small group of female rabbits and then compared their vaginal and clitoral blood flow before and afterward. As theorized, the animals with the atherosclerosis had a lukewarm response when the researchers stimulated their bunny rabbit genitals.

Along these same lines, exercise has been shown to

Chemical analyses of female "ejaculate" have differed. Two labs concluded the samples were the same as urine; four found significant differences. At least one reported the presence of PAP, an enzyme found in the prostatic component of semen. It *is* true that women have a vestigial prostate, a scattering of ducts and glandular material surrounding the urethra, in the G-spot vicinity on the vagina's front wall. It's possible some women expel urine, others a prostatic fluid, and some a mixture. The debate drags on.

*Women get them too. It took science a while to figure this out. You couldn't simply hire a grad student to sit and watch all night, as was done in 1944, when science first confirmed that men were dependably getting erect as they slept (three to five times per night). And the average clitoris was too small for early-model strain gauges. What you needed were some congenitally enlarged clitorises, a fancy strain gauge, and a powerful need to know. In 1970, a trio of University of Florida researchers pulled together all these things and assembled them in a sleep lab. The women in the study were found to have a similar number of erections as did a control group of men, and, as with the men, they usually happened during REM sleep.

improve a woman's ability to get aroused. Which makes sense: Exercise makes the body more efficient at pumping blood. "So when you get into a sexual situation," says Cindy Meston, who ran the study, "the response is both quicker and more intense." (Though it's also possible, Meston allows, that women in better shape simply feel less self-conscious. With less attention devoted to worrying about what their body looks like, there's more to apply to the sensations of arousal.)

If clitoral erectile insufficiency is for real, then you might imagine that the same remedies that work for erectile dysfunction in men might work for women (and, heck, rabbits). And you would not be alone. In the wake of Viagra's monster success in treating ED, Pfizer turned its gaze to women. Partnering with urologists and sex researchers around the country, the pharmaceutical behemoth commenced a massive research venture to see whether genital blood flow was as critical to women's sexual well-being as it is to men's. Or, on a cruder level, to see if Viagra could be marketed to the other half of the planet.

Eight years and 3,000 subjects later, the answer appeared to be no. Viagra did in fact increase blood volume in the nethers, but most women seemed not to notice it. The researchers confirmed what most of them suspected all along: that women's arousal, much more so than men's, rests in the psychological as well as the physiological. And that is why a visit to Cindy Meston's Female Sexual Psychophysiology Laboratory will be coming up.

Before we get there, I feel I must address the female penis pump: as of this writing, the only Food and Drug Administration–approved treatment for female sexual arousal disorder.

The Eros Clitoral Therapy Device is its formal name. Available by prescription, it consists of a small motor inside a plastic housing the size of a bar of soap. Attached to this is a clear flexible cup that fits over the clitoris. Switch on the motor, and you have a handheld suction device. The idea behind a male penis pump is, of course, to pull blood into a limp penis (and keep it there by means of a stretchable band placed around the organ's base). You are making it stiff enough for sex. To get inside a lubricated vagina, a penis needs to be hard enough to push against the opening with one to two pounds per square inch of force.* That is approximately the amount of force required to open a swinging kitchen door. A woman does not need to penetrate anything with her clitoris, any more than a man needs to open kitchen doors with his penis. So why would she use a clitoris pump?

My apologies in advance, but this was something that begged checking into.

---

*We have three Houston researchers to thank for this statistic. In 1985, the trio attached a pressure gauge to the tip of a penis-shaped Plexiglas rod and penetrated a small group of female volunteers. It seems to me that if they wanted to approximate the surface friction that exists in real intercourse, slippery-smooth Plexiglas was a poor stand-in for penis skin. Though I suppose that when you are doing an experiment that involves penetrating coeds in your lab, surface friction is less of a concern than, say, human subjects review board friction.

10

The Prescription-Strength Vibrator

Masturbating for Health

If you call the offices of NuGyn, in Spring Lake Park, Minnesota, there is a good chance you will get Curt Olson on the phone. Olson is not a receptionist. He is the coinventor of the Eros Clitoral Therapy Device. He answers the phone because it's a small company and because he enjoys chatting with folks. While Curt was chatting with me, he told me stories about some of the odder phone calls he's gotten. At the end of each story, he'd pause, and then he'd say, "I'n' that something?" Every now and then, he said, women call to ask where their clitoris is.* "They're pumping on something else. It's like, holy smokes, people!"

---

*Research suggests that these women are rare. In a study of genital self-assessment, fifty women were asked to estimate the size of their clitoris. One of the options was "I cannot locate my clitoris." Happily, no one checked that box.

I asked Curt a question that, surprisingly, he had not been asked before: What made him think that a female penis pump was something the world needed?

"Well," said Curt. "One day my boss and I were making a list of what we could do for our next product." Uro-Metrics—which owns NuGyn—makes diagnostic devices for male erectile dysfunction. The Eros is their first excursion into female sexual dysfunction, and it appeared at first blush that they were not fully appreciating the difference. "We just saw the void: The penis pump was for the men and there wasn't anything for the women."

"But Curt," I said, "women don't need erections to have sex. So why would they need this?"

Curt replied that it wasn't about erections. "It's increased blood flow that brings about the orgasm, so what better to do than pump it? Increase that blood flow." So was this something you'd use to prime yourself before intercourse, much as you might use a vibrator? Or was this something whose regular use would somehow alter your physiology and render you permanently more primed— something that would make it easier for you to become aroused even when the Eros was tucked away in its little satin pouch? In other words, a cure for FSAD. Presumably, that is what they're shooting for.

Curt said that yes, they were after "more of a physical change rather than just a stimulation." He suggested that routinely pulling more blood to the area might help clear up fibrosis in the erectile tissue, which does contribute to erectile dysfunction in men. Jennifer Berman, a Los Angeles urologist and TV sex expert who was involved in some of the Eros clinical trials, also had the impression that it was functioning more as a long-term therapy. "It's something you'd use independent of sex," she said. "Like doing your push-ups and your jumping jacks, you would use your Eros device as well."

I asked Curt if I could borrow an Eros to see what it does. "You want to borrow one?" He seemed unenchanted. "How about if we don't want it back? How about if you just keep it."

The directions tell you to use the Eros for one minute, rest for a minute, and then use it for another minute. And to repeat this cycle three to five times. The Eros is a big, fat tease. I am here to tell you that anyone who makes it through one or two start-stop cycles very quickly loses interest in watching the secondhand and keeping track of which cycle she's on and when to rest. The Eros will turn you into a masturbatory layabout. But does it improve the sex you have with parties other than your Eros device?

In a 2002 paper, women's scores on the Female Sexual Function Index improved significantly after three months of Eros therapy. This was true for women with FSD and for controls with no sexual complaints. However, the Female Sexual Function Index includes a lot of questions along the lines of: Over the past four weeks, how often did you feel sexually aroused during sexual activity or intercourse? How often did you become lubricated during sexual activity or intercourse? When you had sexual stimulation or intercourse, how often did you reach orgasm? Well, if the woman includes her Eros encounters as part of that sexual activity, and she's using the thing four times a week, as the therapy calls for, of course her score is going to be higher.

Despite the shortcomings of the studies, given that there are no side effects, that the "therapy" amounts to near-daily doses of self-pleasure, it is hard to make a case against giving it a try.

Though there's that pesky $400 price tag. Why do women need to spend $400 for suction?

"Curt?"

"Yes, ma'am."

"Could you just use a small vacuum cleaner?" Curt surmised that that would likely cause some bruising. Possibly, it would cause more than that. For the Florida man who was found slumped on his dining room table after neighbors reported hearing "a vacuum cleaner running continuously for a long time," it caused a fatal heart attack and thermal burns on "areas in direct contact with the beater bar." The *American Journal of Forensic Medicine and Pathology* case report★ "Vacuum Cleaner Use in Autoerotic Death" includes a photograph of the deceased slumped over his vacuum, a seventies-era upright, with one arm encircling the canister in the manner of a lover's embrace. (It was not their first encounter. The man's wife "had surprised him masturbating with a vacuum cleaner" once before, though surely not as much as he had surprised her.)

Any kind of arousal will draw more blood to the area. What if sexually dysfunctional women were to use a garden-variety vibrator four times a week? Or how about just their finger? Are we sitting on (whee!) a cheap, simple, safe, universally available, highly pleasurable treatment for female

---

★What is it with pathology journals and autoerotic deaths? Every other issue seems to have a case report of some heedless, autoasphyxiated corpse with ill-fitting briefs and a black bar across his eyes. Occasionally, they seem to be in there for sheer color, as in the case of the young Australian who perished from "inhalation of a zucchini." This one raises more questions than it answers. Was he trying to intensify his climax by vegetally choking himself, or was it a case of overexuberant mock fellatio? (We do learn that the zucchini was from his wife's garden, admittedly a nice touch.)

sexual dysfunction? I called Jennifer Berman. "Hm," she said. "I mean, technically I guess you could say that. Whether the Eros device is any better than masturbating, that I can't answer."

I emailed Arno Mundt, a University of Chicago professor of gynecologic oncology whose name is on a different Eros therapy study. "Dear Dr. Mundt," I wrote. "If bringing blood to the clitoris/vulva more often (with the Eros) helps with arousal, lubrication, orgasm, etc., then would simply masturbating, with a vibrator or manually, four times a week also help?" (In retrospect, a line or two of introductory chitchat was in order.)

"Good question," came the reply. "I will defer this to Maryann."

Maryann Schroder, a licensed sexologist at the University of Chicago Hospitals, is the principal investigator on the study.

"You have posed a very interesting question," she said. "It hasn't been studied, if you can believe." She reminded me of what happened to the last person who got involved with masturbation as a beneficial activity: Surgeon General Joycelyn Elders. Former President Bill Clinton dismissed Elders after she suggested, in a World AIDS Day speech, that masturbation was something that "should perhaps be taught."

"Can you imagine if I tried to get funding for a study that had masturbation in the title?" And then, quite unintentionally, Dr. Schroder delivered the ultimate masturbation-research sound bite. "Masturbation," she said, "is a touchy area."

Not everyone who deals with masturbation on a professional level has to concern themselves with what the government thinks. Not everyone gets their funding from research grants. Some masturbation professionals get their funding from the sales of Vibrating Port-A-Pussies and Mr. Fred Jelly Dongs. I have made an appointment to visit Marty Tucker, chairman and founder of the world's second-largest sex-toy manufacturer. I came across Marty's name on U.S. Patent 5,693,002: Sexual Appliance Having a Suction Device Which Provides Stimulation. Maybe Marty can answer my question about the medical benefits of regular self-stimulation, suction-based or otherwise.

there are images that stay with you your whole life, whether you want them to or not. Here is one that I imagine will make the cut. A man in a blue smock and a hairnet walks across a factory floor with an armload of enormous chocolate-brown dildos. He is loaded down to the point of absurdity. He is Audrey Hepburn leaving Bergdorf's in some 1960s romantic comedy, her arms piled so high with packages that she can barely see over the top. I want to trip him, not out of meanness, but just to see the penises fly through the air and rain down around us.

Marty Tucker is showing me around the Topco manufacturing floor. It is the size of a football field and as loud as a Super Bowl. Marty is yelling over the noise.

"THESE ARE VAGINAS." He makes a sweeping motion with one arm, drawing my gaze to a long, narrow surface heaped with objects that do not suggest any facet of human anatomy.

"VAGINAS?"

"IT'S A VAGINAL PRODUCT. A MASTURBATION TOY." He picks one up. "THIS IS A TUBE THAT YOU WOULD ENTER FROM THIS SIDE AND YOU WOULD BE INSIDE OF THIS HERE." He is using "you" in the sense of *your penis, were you the sort of person who (a) had a penis and (b) shopped at places like Topco.* "THERE'S A

GEAR SYSTEM THAT'S ROTATING A PIECE WITH
BEADS ON IT, TO HIT THE FRENULUM.*..."

Marty talks about his goods frankly and technically,
as though they were car parts or kitchen appliances. He
said later, "It's just product. Everybody who works here is
immune to it. It's not sexual anymore, it's like a key chain
or a wallet. It's nothing."

Now we have paused to watch a team of women, wearing latex gloves, whose job is to rub a light film of red paint
into the testicles and glans of large pink dildos, to pinken
them, "to give them the realism." Realism strikes me as
an odd choice for this product. These are huge phalluses,
comically huge, with veins like jungle vines. The women
are chatting and laughing while they work. Their movements are inadvertently erotic; the hand-staining of a
dildo tip could be the efficient caress of a sex worker. The
women are Latinas in their thirties and forties, as are many
of Tucker's employees here. If you went out to the parking lot, you would find rosaries and Virgin Marys hanging from the rearview mirrors. I ask Marty if the women's
families know what they do for a living.

Marty quiets his voice. "My experience is that they
really don't talk about it—the fact that they're working
with a ten-inch penis. There's one area in here where the
girls are sewing hair onto the vaginas for pubic hair. I asked
a girl one day, 'Do your parents know what you do?' She
says, 'No, I just tell them I work in plastics.'"

Marty steers us to a small conference room, where
it's quieter and we can talk. The walls are bare except for
six 8-by-10 aerial photographs of the Topco building, a

*The nerve-dense bit of tissue on the underside of the penis, where shaft
meets glans. "Along with the tip and the testicles, these are the sensitive
parts," says Marty. "The whole rest of the penis, you could throw away."

squat, anonymous-looking 12,000-square-foot chunk of Chatsworth, California, business park. Tucker himself looks very much like what he is: the chairman of a successful multinational manufacturing firm. His cuffs are monogrammed so ornately as to be unreadable. He wears an unsubtle diamond ring on each of his ring fingers, a navy blue suit, and a tie with a travel motif of foreign flags. A New York accent grabs hold of his words.

"So what can I do for you?"

I had sent Marty a detailed email asking about the relative merits of clitoral suction and vibration and about whether both of these, regularly practiced, could improve a woman's responsiveness. Yet somehow I have given him the impression that I have come to write about him. And possibly I should have.

I repeat my intended mission. Marty listens. "Okay. Suction will pull more blood and make the clitoris more sensitive. That's what suction does." He refers to the sex toy whose patent I mentioned, which turns out to be very simple: a ring that encircles either a real penis or a phallic sex toy and, attached to it, a suction cup. "So the ring can be worn by the guy, and when he's inside of his partner, the suction cup is against the clitoris." Marty pronounces it cli-TOR-is, rhymes with Lavoris.* The item is no longer sold.

I ask Marty if he has any reason to believe that regularly

---

*Merriam-Webster.com's preferred pronunciation is CLIT-oris. If you click on the little speaker icon, you can hear a nice lady saying "CLIT-oris" out loud for you, over and over, as many times as you click. The nice lady will also say "cervix" and "nipple," but it is the nice man who gets to say "vagina," "vulva," and "orgasm," plus all the male genital words. Smelling sexism, I entered "housewife," which was read aloud by the woman, as was "maid," "stewardess," and "flower." However, it is also the woman who pronounces "linebacker," "doctor," "president," and "fireman." So never mind. Can you say "waste of half an hour"?

bringing more blood to the crotchal area—with suction, with vibration, with any sort of masturbation—would improve a woman's responsiveness or—Word of the Hour—orgasmicity. It's a fine opportunity for him to promote sex toys as therapeutic devices, but he doesn't take the bait. He says he hasn't heard anything to that effect.

On the way back to the lobby, we stop on the factory floor again, in the baking area, where the molds in their plaster casts go in and out of ovens. The countertop is Pollacked with drips and spills of liquid plastic. The noise is deafening: clankings, thrummings, pneumatic exhalations.

"HERE YOU HAVE AN ANUS." Marty says it's a model of a porno actor. Porn stars come in to Topco, to a special room, where the staff make plaster casts of their penetratable regions. This includes their faces, which can be purchased separately or put on a doll body. The top stars get a royalty for each orifice sold.

Marty's hand rests on a model that is cooling in its mold, buttocks-down. "YOU CAN FEEL THAT THE MATERIAL IS STILL WARM RIGHT NOW, IT'S STILL SOFT." From the back, it looks innocent, edible, like a chocolate dessert product. It's all I can do not to take a bite.

masturbation therapy for women is not altogether new. It is, in fact, altogether old. Genital massage was a common medical treatment for sexually frustrated women as far back as Hippocrates' day. The Hippocratic physician, of course, lacking batteries and Topco catalogues, had to make do with his fingers (or, often, those of a midwife).

For centuries, medical texts included long discussions of a condition called hysteria, a sort of vaguely defined sexual dysfunction based on spectacular misrepresentations of female anatomy and sexuality, and treated by, among other things, manual manipulations. The ancient Greeks, as we've learned, thought that women produced their own semen, released at the climax of intercourse, and that the mingling of male and female seed formed the basis of conception. Young widows, with no sexual outlet and a consequent log jam of womanly seed, were said to be especially prone to hysteria—or "womb fury." (The widower was spared because he regularly jettisoned some during nocturnal emissions.) The notion persisted for centuries. Audrey Eccles quotes a physician in *Obstetrics and Gynaecology in Tudor and Stuart England:* "It is most commonly the widowes disease; . . . when the seed is thus retained it corrupts, and sends up filthy vapours to the brain." A typographically deranged colleague named Maubray concurred: "By a long *Detention* there, [the seed] may be converted into VENOM, or a *Poysonous Humour. . . .*"

The cure, logically enough, was to contrive a climax. Though no one came right out and said that that's what he was up to. Chapter LXVIII of *Aetios of Amida: The Gynaecology and Obstetrics of the VIth Century,* A.D. outlines tactics for triggering the release of she-semen. "The midwife having taken [various oils] with her fingers, she should . . . rub the part gently and for a long time. . . ." Eventually, "much thick and viscid sperm [was] expelled, and the woman was freed without delay from her distressing affliction." Presumably, Aetios was mistaking vaginal lubrication for semen. Gynecology was but a sideline interest for Aetios (best known for his eye, ear, and nose texts), and it showed. Women who came to him for contraceptive advice were told to wear a piece of cat liver in an ivory tube attached to their left foot. Though I suppose this might well keep you from getting pregnant, in the same way that wearing Crocs might.

While awkward to be sure, genital manipulation was preferable to hysteria's other treatment: the evil smell. This line of attack was based on the belief that hysteria was associated with a retracted uterus; foul odors were inhaled to repel the uterus, in the hope that it would retreat back down the body cavity into its rightful position. For ten-plus centuries, the womb was considered less an organ than an independent creature, able to move about the woman's body like a badger in its den. Aetios of Amida prescribed the following: "Place at the nostrils a pot of stale urine." *Soranus' Gynecology* describes anointing the patient's nose with "squashed bed bugs." From the Tudor era, we hear that "also highly esteemed was a fume made of 'the warts which grow upon Horses Legs. . . .'" Overall, it was hard to escape the suspicion that the early gynecologist was not the caring and supportive creature that she is today.

Evil odors came and went, but "pelvic massage" for hysteria persisted all the way through the Victorian era to the first half of the twentieth century. The earliest vibrators weren't being sold to women; they were being sold to physicians to make their job easier. Depending on the practitioner's skills and the woman's inhibitions, manually instigating climax in a doctor's office could take upward of half an hour. The vibrator was a godsend, reducing the chore to a few minutes.

Rachel P. Maines, the sort of historian the world needs more of, wrote a book on this topic. *The Technology of Orgasm* is packed with amazing information, but none more so than this: "There is no evidence that male physicians enjoyed providing pelvic massage treatments. . . ." It was, she said (and sort of still often is) "the job nobody wanted." I had imagined doctors getting caught up in, and turned on by, their patients' reactions. But Maines found no evidence of this. She states that most of these physicians did not even understand that the climax of the treatment they were providing was an orgasm.

When vibrator manufacturers finally came out with home models, the ads were predictably opaque. While some made oblique references to the devices' true charms ("makes you fairly tingle with the joy of living"), most dispersed a smoke screen of vague health claims. Others ventured deep into the ludicrous. Maines's book includes an ad from 1916 showing a woman with a vibrator held up to her cheek, the caption claiming that the device would "bring social and business success." A pair of Star Vibrators were advertised in 1922 as "Such Delightful Companions! . . . Perfect for weekend trips," as though they could serve up witty repartee and spell you at the wheel.

Even today, vibrators are sold as "massagers" to women who are uncomfortable buying sex toys. The small appliance company Wahl, for instance, sells a trio of massagers on its Web site with no explanation of what they're for. (The fact that nitetimetoys.com is the first listing that comes up when you Google "Wahl massagers" provides a hint.)

The company bio of the late John Wahl notes that he served not only as Wahl president but also in a leadership capacity at St. Mary's Catholic Church. And that his brother Raymond Wahl is a monsignor. I'm not saying there's a link between Catholicism and sex toys. I'm just saying I've got a brand-new interpretation of Isaiah 49:2 ("The Lord . . . hath made me a polished shaft").

Why weren't hysteria sufferers simply told to go home and masturbate twice a week? Because, as you will recall from chapter 6, masturbation has a long history as a shameful, dangerous, and much-discouraged act.

But now that we all know better, should gynecologists be recommending masturbation as a treatment for sexual dysfunction? Are orgasms the ticket to sexual health? I called Cindy Meston, whose laboratory we are headed for shortly. Her answer was yes. Her graduate student Lisa Dawn Hamilton recently completed a study that tracked the testosterone levels of women in long-distance relationships. (Testosterone is the hormone most closely linked to sexual desire, and is sometimes prescribed to women who complain of a low libido.) Testosterone levels were significantly higher when the women were having sex, as compared to the days when their partners weren't there. (The participants promised not to masturbate for the duration of the study.) "It's looking like sex in and of itself can be therapeutic," says Meston. "It makes you enjoy sex more and want to have sex more. I think the whole use-it-or-lose-it thing definitely applies to women."

Meston agreed that a $25 Micro Tingler from Marty Tucker's warehouse probably affords much the same benefits as a $400 Eros Therapy Device. However, she made the point that there are women out there who would be uncomfortable with a treatment that consists of masturbation, with or without a sex toy—not to mention doctors who are uncomfortable prescribing it. For them, as Meston says, "the guise of it being an FDA-approved medical device takes some of the taboo out of it."

The taboo issue might also explain the impressive sales records of some of the quack powders and oils sold online as arousal boosters for women. While they sometimes contain spices or chile extracts that create a mild tingling sensation, the key ingredient, Meston says, is more often one's own hand. "They come with these instructions like, 'Apply to clitoris and labia and rub *really well* for an extended period of time. Make sure you rub *really, really well. . . .*' "

In 1999, somewhere in the state of Israel, a man began hiccuping and could not stop. He tried the silly things his friends suggested. He pulled on his tongue and rubbed the roof of his mouth with a Q-tip. He tried chlorpromazine, metoclopramide, defoaming antiflatulents even. Nothing worked. The man grew increasingly anxious. He could not sleep or concentrate on his work. On the fourth day, still hiccuping, the man had sex with his wife. His condition persisted all the way through the act, and then, once he ejaculated, the hiccups stopped. *Canadian Family Physician* published a case report about the man, under the title "Sexual Intercourse as a Potential Treatment for Intractable Hiccups." Unattached hiccuppers were advised that "masturbation might be tried."*

Are there other nonsexual health benefits to be derived from orgasm? Affirmative, say Rutgers University sex researchers Barry Komisaruk and Beverly Whipple. Their readable and comprehensive *The Science of Orgasm* says that people who have regular orgasms seem to have less stress and enjoy lower rates of heart disease, breast cancer, prostate cancer, and endometriosis.

They also appear to live longer. British researcher G. Davey Smith and two colleagues calculated that over a span of ten years the risk of death among men who had two or

---

*Or, if they are followers of sixteenth-century naturalist Li Shih-chen, sun-dried, powdered wolf epiglottis. Li's hiccup remedy, found in the *Chinese Materia Medica*, is probably quite effective, for in the time it takes to track a wolf and sun-dry its epiglottis, even the most stubborn case of hiccups will invariably have passed.

more orgasms a week was 50 percent lower than among those who had them less than once a month. (Obviously, the researchers had to control for factors like social class, smoking, and age.) Catholic priests, as compared with their noncelibate Protestant counterparts, have higher rates of early death. This last bit was reported—though without a source—in a 1990 *Sports Medicine* article entitled "The Sexual Response as Exercise." The author, a psychologist named Dorcus Butt,★ then at the University of British Columbia, states that the muscle tone, strength, and straining involved in orgasm are similar to that of "jumping, gymnastics, tennis, football. . . ." Yet one more reason the Catholic Church should condone sex, or jumping, among its clergy.

Orgasm may be, as Butt says, "the most basic form of physical exercise," but that doesn't mean sex is a particularly good workout. In 1984, psychiatrist Joseph Bohlen brought ten married couples into a laboratory at Southern Illinois University School of Medicine and measured the men's heart rate, metabolic expenditure, and oxygen uptake during five different sexual activities: foreplay, intercourse (once in the missionary position, once with the wife on top), fellatio, and masturbation. Bohlen concluded that sex was, at best, "light to moderate" exercise of short duration. However, given that "the mask used to collect the husband's expired air kept him from kissing . . . , and the ECG electrodes and blood pressure cuff hoses restricted body movement," it is possible that the sex being had in Dr. Bohlen's lab was less exuberant than usual.

People with spinal cord injuries may derive a unique benefit from orgasm. If you are paralyzed, say, or you have

---

★I can't speak for Butt, but Dorcus was once a popular and desirable name. There was a magazine for "embroiderers and needle artisans," popular in the early twentieth century, titled *Dorcus*.

multiple sclerosis, you may find that orgasm relieves you of the leg stiffness and muscle spasms collectively known as spasticity. Alfred Kinsey noticed this during his attic observations of men with cerebral palsy. Apparently, the benefit lingers for some time afterward. Researchers have found that a session with a rectal probe electroejaculator dampens leg spasticity for, on average, eight hours.

You may be curious as to who got the idea to look into this. The rectal electroejaculator★ is, after all, a device intended for use on livestock. Artificial inseminators electroejaculate bulls and stallions to obtain the semen used to artificially impregnate cows and mares. Men with spinal cord injuries—who often can't ejaculate in the usual manner—have themselves electroejaculated, in fact, for similar reasons. It all began in 1948. A team of doctors at the Cushing Veterans Administration Hospital in Massachusetts, hoping to obtain sperm that could be used to impregnate the wives of paralyzed veterans, revved up a McIntosh No. 5005 portable, wall-mounted electrophysiotherapy machine. (Electricity was a fad health treatment popular from the late 1800s to mid 1940s. Rachel Maines describes

---

★Far from the worse thing that ends up in rectums. "Rectal Foreign Bodies: Case Reports and a Comprehensive Review of the World's Literature" includes a list of objects doctors have removed from rectums over the years. Highlights: a frozen pig tail (one of the 7 female cases in the total caseload of 202), a bottle of Impulse Body Spray ("incarcerated" in a thirty-seven-year-old lawyer), a parsnip, a plantain (with condom), a dull knife, a cattle horn, a salami, a jeweler's saw, and a plastic spatula. Multiple holdings in the same rectum are listed under the heading "Collections." These include several that could pass as still-life titles ("oil can with potato," "2 apples," "402 stones"), several that probably couldn't ("umbrella handle and enema tubing," "lemon and cold cream jar") and one that suggests a quiet evening in the Biltmore ("spectacles, suitcase key, tobacco pouch, and magazine").

an advertisement for one such device, showing "electrodes for every conceivable bodily orifice." Presumably, the Cushing team was aware that ejaculation was a common side effect of rectally administered electrotherapy.)

Alas, though the Cushing doctors obtained* semen from all eighteen men, only two of the wives became pregnant. (Partly because when the cremaster muscle is paralyzed, the testes can't be lowered away from the body to cool, and the sperm overheat.)

In 1981, British sex researcher Giles Brindley set out to hone the craft of electroejaculation for fertility purposes. To figure out which nerve fibers were most expeditious, he first experimented on baboons and rhesus monkeys. As the electrotherapy fad had long ago faded, Brindley devised a homemade ejaculator, consisting of an electrode mounted on the tip of his gloved finger. Presently the baboons were excused and Brindley attempted some "experiments on myself."† Because Brindley's spinal cord is intact, he could

---

*"Obtained" being a handy euphemism. Paralysis disables the sphincter that normally closes to keep semen from heading off into the bladder. Thus, electrically prompted ejaculations are often "retrograde," disappearing into the manly recesses unless someone "milks" the semen out into the light of day.

†A Brindley tradition. At a 1983 urology conference, Brindley delivered a lecture about a new impotency drug, papaverine, that produced robust erections when injected directly into the penis. He began by showing his audience, a group of around eighty urologists and their wives—many en route to the conference cocktail party and dressed in formal attire—a series of slides of his own penis, after various dosages. He then revealed that, five minutes earlier, he had injected himself with papaverine. He pulled the fabric of his track suit tightly against his hips to reveal the outline of his medicated member. Not satisfied, he then pulled down his pants, revealing, in the words of eyewitness Laurence Klotz, "a long, thin, clearly erect penis. " Klotz's account of the event was published in

feel pain that paralyzed men could not, and he had to stop at less than a quarter of the voltage that would trigger ejaculation.★ Brindley made 256 attempts at electroejaculating 84 men with spinal cord injuries. Fourteen wives were inseminated, but, owing to the inferior quality of the sperm, only one conceived. Some of the wives soldiered on anyway. In Brindley's paper, under the evocative heading "Domestic Electroejaculation," he recounts that ten women had been taught to administer the voltage at home.

In the course of Brindley's study, some of the subjects reported that for several hours afterward, their leg spasms had quieted. This information found its way to veterinarian and livestock electroejaculator designer Steve Seager. Not one to pass up an opportunity for lateral marketing, Seager got a grant to do a formal study of the efficacy of one of his Electrostimulation Units for reducing leg spasticity—in both men and women. It worked. And that is the

BJU International in 2005. "He paused, seeming to ponder his next move. The sense of drama in the room was palpable. He then said, with gravity, 'I'd like to give some of the audience the opportunity to confirm the degree of tumescence.' With his pants at his knees, he waddled down the stairs. . . . As he approached [the audience], erection waggling before him, four or five of the women in the front rows threw their arms up in the air . . . and screamed. . . . The screams seemed to shock Professor Brindley, who rapidly pulled up his trousers . . , and terminated the lecture."

★Which perhaps explains the dearth of ejaculator references on alternative-sex Web sites. I found just one, a reference to the Bailey Ejaculator, in the Yahoo Mechanical Sex newsgroup. "It sounds rather captivating," says the posting. "It sounds awesome moreover. Has anyone got any info?" Ed Lehigh, assistant vice president at Western Instruments, where they make the Bailey, told me he was unaware of a recreational side market for his product, though he recently got an order for three livestock ejaculators from an L.A. porn producer, perhaps seeking to boost productivity among the cast.

story of how rectal probe electrostimulation came into use as a therapy for muscle spasticity in people with spinal cord injuries.

These days there is a cheaper, less intimidating alternative. The technique is called Transcutaneous Mechanical Nerve Stimulation, a.k.a. pressing a vibrator to the underside of your penis. Not just any vibrator, but a high-amplitude one made by FertiCare,* called the Personal Penile Vibrator—or, should you happen to select the Spanish version of FertiCare's Web site, El Vibrador del Pene.

Attentive readers may be thinking: If paraplegics can't feel anything down there, how do they get aroused and have an orgasm? That is one of many mysteries being solved at the University of Alabama School of Medicine's sexual physiology lab. The others are more universal: What exactly *is* an orgasm? Where in the body do you feel it? Can dead people have them too?

---

*"High-amplitude" meaning that the part that vibrates back and forth travels a longer distance in each direction. The FertiCare surpassed both an Oster and a Sunbeam vibrator in "An Analysis of 653 Trials of Penile Vibratory Stimulation in Men With Spinal Cord Injury." It was news to me that either of the aforementioned wholesome small appliance companies made vibrators. They surely do not flaunt them. It is easier to track down an Oster animal nail grinder or an Oster arepa maker than an Oster vibrator.

11

The Immaculate Orgasm

Who Needs Genitals?

marcalee Sipski is an expert in a field with few experts. When I tell you what the field is, you will understand why the experts are scarce. Sipski, a professor at the University of Alabama School of Medicine, is an authority on sexuality among people with spinal cord injuries and diseases. Most people, even most M.D.s, are uncomfortable sitting down with a paraplegic and having a talk about, say, how to have intercourse with a catheter in your penis. Sipski is fine having that talk,★ and she is fine with my coming to her lab while a subject is there.

Very little fazes Dr. Sipski. For her video *Sexuality Reborn: Sexuality Following Spinal Cord Injury*, she managed to recruit four couples to talk frankly (". . . and there's the

---

★"The catheter can be folded back over the penis and both the penis and catheter covered with a condom."

stuffing method") about how they have sex and even to demonstrate on-camera. They participated because they, like Sipski, were aware of the potentially ruinous effects of a spinal cord injury on a couple's sex life and how hard it can be to find doctors willing to address the issue in a constructive, nuts-and-bolts manner.

Sex research is a relatively recent development in Sipski's career. For years, she maintained a private practice in rehabilitation medicine. (Christopher Reeve was one of Sipski's patients, as was Ben Vereen.)* Over time, she grew curious about the surprisingly high percentage of patients who said they were still able to have orgasms. For decades, the medical community—being for the most part non-disabled—had assumed that people with para- and quadriplegias couldn't have them. It was a logical assumption: If a person's spinal cord is broken at a point higher than the point at which nerves from the genitals feed into the spine, then there should be no way for the nerve impulses to make their way past the injury and up to the brain. And thus, it was further assumed, no way for the person to reach orgasm.

Yet 40 to 50 percent of these men and women, according to several large surveys, do. Sipski decided to investigate. She recruited people with all different degrees and levels of spinal cord injuries for a series of studies, to see if she could find any patterns.

People with spinal cord injuries provide a unique window onto the workings of human orgasm. If you examine lots of people—some whose injuries are high on the spine,

---

*Vereen came in for rehabilitation after he was struck by a car (though not paralyzed) while walking on the Pacific Coast Highway some years back. Sipski recruited Vereen to do the introduction on *Sexuality Reborn*, which he undertook with admirable dignity and no dancing.

some down low, some in between—you can eventually isolate the segments of the nervous system that are crucial to orgasm. You can begin to define what exactly an orgasm is. (A recent review of the topic listed more than twenty competing definitions.) Once you have an accurate definition of what orgasm is and how it happens, then you will, hopefully, have some insight into why it sometimes doesn't. Studying people with spinal cord injuries might benefit the non-disabled as well.

It is a testament to Sipski's reputation in the disabled community that more than a hundred men and women with spinal cord injuries have traveled to her lab to be part of a study. Unless you are extremely comfortable with your sexuality, masturbating to orgasm in a lab while hooked up to a heart-rate and blood-pressure monitor is, at best, an awkward proposition. It's even more daunting when you have a spinal cord injury: Among those who can reach orgasm, it takes on average about twice as long to get there. Though Sipski's subjects are alone behind a closed door, they can hear voices and sounds on the other side of the wall. They can tell that people are out there, timing them, monitoring them, waiting for them to finish.

The people out there this morning are uncommonly disruptive. This is because one of them is me, and because Sipski's colleague Paula Spath said that by climbing up onto her desk and pressing my nose up to the one-way glass, I could get a peek at the experimental setup. I have on a skirt that does not lend itself to scaling office furniture. I lost my balance and crashed into Paula's monitor, which slid across the computer it was standing on, knocking off a row of knickknacks and causing Paula to leap back and let out the sort of high-pitched exclamation that might more appropriately be heard on the yonder side of the wall. It's a wonder anyone invites me anywhere.

A woman I'll call Gwen is under the covers inside the lab. Aside from a caddy in the corner that holds the physiological-monitoring equipment, the lab resembles a scaled-down hotel room: there is a bed with a tasteful bedspread and extraneous throw pillows, a chair, a bedside table, a framed art print, and a TV for viewing erotic videos. Helping Gwen with her assignment is an Eroscillator 2 Plus, a vibrator endorsed by Dr. Ruth Westheimer and developed by Dr. Philippe Woog, the inventor of the first electric toothbrush.*

While Gwen eroscillates, Sipski explains what transpired before I arrived. All her subjects are given a physical examination to determine the extent and effects of their injury and its precise location in the spinal cord. One theory held that the people who could still have orgasms were those whose injuries were incomplete—meaning the spinal cord wasn't completely severed and that some of the nerve impulses from the genitals were squeaking through and

---

*For years, Dr. Woog had been aware that women were using his invention as a vibrator. Every now and then, returned toothbrushes that passed through quality control "clearly seemed to have been used in that way," said his son Lionel Woog, who oversees marketing for the vibrator company Advance Response. At a certain point, a lightbulb went on in the elder Woog's head, and he set to work on a vibrator. "It's the same idea," said Lionel. "You want to stimulate the tissue without damaging it." Lionel told me the story of the Inuit, and how their gums deteriorated when they moved to settlements and began eating processed foods instead of raw animal parts. Vigorous chewing, he explained, stimulates bone growth and keeps gums healthy. Woog's Broxodent electric toothbrush used to be given to crews on nuclear submarines. "They ate a lot of canned foods," said Woog, and the toothbrush helped keep their gums in shape. It was a popular item with the men, maybe even for that reason. As Lionel Woog says, "You have to masticate."

reaching the brain. Another possibility was that the orgasmic ones were those whose breaks were below the point where the genital nerves feed into the spinal cord.

It turned out that while both these things can make a difference, neither was an ironclad deciding factor for orgasmicity. People with high spinal cord injuries could have them, and so could some with complete spinal cord injuries. Based on Sipski's data, only one thing definitively precludes orgasm: a complete injury to the sacral nerve roots at the base of the spine. Injuries here interfere with something called the sacral reflex arc, best known for its starring role in bowel and bladder function. The sacral reflex arc is part of the autonomic nervous system, the system that controls the workings of our internal organs. "Autonomic" means involuntary, beyond conscious control. The speed at which the heart beats, the peristaltic movements of the digestive system, breathing, and, to a certain extent, sexual responses, are all under autonomic control.

Sipski explains that when you damage your spinal cord, you primarily block the pathways of the somatic, not the autonomic, nervous system. Somatic nerves transmit skin sensations and willful movements of the muscles, and they travel in the spinal cord. But the nerves of the autonomic nervous system are more complicated, and not all of them run exclusively through the spinal column. The vagus nerve, for example, feeds directly from the viscera into the brain; Rutgers University researchers Barry Komisaruk and Beverly Whipple have posited that the vagus actually reaches as far down as the cervix, and that that may explain how people with spinal cord injuries feel orgasm. Either way, autonomic nerves seem to be the answer to why quadri- and paraplegics can often feel internal sensations—menstrual cramps, bowel activity, the pain of appendicitis. And orgasm.

"Think about it," Sipski is saying. "Orgasm is a not a surface sensation, it's an internal sensation." Sipski routinely asks her spinal-cord-injured subjects where they stimulated themselves and where they felt the orgasm. Of nineteen women who stimulated themselves clitorally, only one reported that she'd felt the orgasm just in her clitoris. The rest ran an anatomical gamut: "bottom of stomach to toes," "head," "through vagina and legs," "all over," "from waist down," "stomach first, breast tingle, then vaginally."

It is strange to think of orgasm as a reflex, something dependably triggered, like a knee jerk. Sipski assures me that psychological factors also hold sway. Just as emotions affect heart rate and digestion, they also influence sexual response. Sipski defines orgasm as a reflex of the autonomic nervous system that can be either facilitated or inhibited by cerebral input (thoughts and feelings).

The sacral reflex definition fits nicely with something I stumbled upon in the United States Patent Office Web site: Patent 3,941,136, a method for "artificially inducing urination, defecation, or sexual excitation" by applying electrodes to "the sacral region on opposite sides of the spine." The patent holder intended the method to help not only people with spinal cord injuries but those with erectile dysfunction or constipation.

Best be careful, though. The nervous system can't always be trusted to keep things straight. *BJU International* tells the tale of a man who visited his doctor seeking advice about "defecation-induced orgasm." For the first ten years, the paper explains, he had enjoyed his secret neurological quirk, but he was seventy now, and it was wearing him out. Horridly, the inverse condition also exists. Orgasm-induced defecation was noted by Alfred Kinsey to afflict "an occasional individual."

The electronics term for circuitry mix-ups is crosstalk:

a signal traveling along one circuit strays from its appointed route and creates an unexpected effect along a neighboring circuit. Crosstalk explains the faint voices from someone else's conversation in the background of a telephone call. Crosstalk in the human nervous system explains not only the man who enjoyed his toilette, but also why heart attack pain is sometimes felt in the arm, and why the sensations of childbirth have been known to include orgasmic feelings. Orgasms from nursing (or nipple foreplay) are another example of crosstalk. The same group of neurons in the brain receive sensory input both from the nipples and the genitals. They're the feel-good neurons: the ones involved in the secretion of oxytocin, the "joy hormone." (Oxytocin is involved in both orgasm and the milk-letdown reflex in nursing mothers.)

here is something eerie about spinal reflexes: You don't need a brain. For proof of this, you need look no further than the chicken that sprints across the barnyard after its head is lopped off. Eerier still, you don't even need to be alive. The spinal reflex known as the Lazarus sign has been spooking doctors for centuries. If you trigger the right spot on the spinal cord of a freshly dead body or a beating-heart cadaver—meaning someone brain-dead but breathing via a respirator, pending the removal of organs for transplant—it will stretch out its arms and then raise them up and cross them over its chest.

How often do the dead move? A research team in Turkey, experimenting on brain-dead patients at Akdeniz University Hospital over a span of three years, were able to trigger spinal movement reflexes in 13 percent of them. (In a Korean study two years later, the figure was 19 percent.) Most of the time, the dead just jerk their fingers and toes or stretch their arms or feet, but two of the Turkish cadavers were inspired to perform the Lazarus sign.

Reflexive movements can be extremely disquieting to the medical professionals in the OR during organ procurement surgery—so much so that there was a push in England, around 2000, to require that anesthesia be given to beating-heart cadavers. New York lawyer-physician Stephanie Mann, who publishes frequently on the ethics of brain death and non-responsive states, told me that although beating-heart cadavers may appear to be in pain, they are not. "Certainly not in the way you and I perceive pain. I think the anesthesia is administered more for the doctors' discomfort than for the cadaver's."

Mann said—because I asked her—that it might be possible for a beating-heart cadaver to have an orgasm. "If the spinal cord is being oxygenated, the sacral nerves are getting oxygen, and you apply a stimulus appropriately, is it conceivable? Yes. Though they wouldn't feel it."

I tell Sipski she should do a study.

"*You* get the human subjects committee approval for that one."

Okay!" It's Gwen's voice over the intercom. "I'm finished." She has a soft, swaying Alabama accent, "okay" pronounced UH-KAI. Paula tells Gwen to lie quietly for a few minutes and watches the monitor. She is looking for the abrupt drop-off in heart rate and blood pressure that signals that an orgasm has come and gone.

Gwen has agreed to talk with us for a few minutes before she leaves. She sits in a chair and looks at us calmly. If you did not know what she had been doing, you would not guess. Her hair is neat and her clothes are unrumpled.

Only her heart rate as the experiment began (117 beats per minute) betrayed her unease.

Gwen was diagnosed with multiple sclerosis in 1999. (Sipski began collecting orgasm and arousal data on MS patients earlier this year.) Her beauty and poise belie the seriousness of her condition. She says she is tired all the time, and her joints hurt. Her hands and feet sometimes tingle, sometimes go numb. She has trouble telling hot from cold and must have her husband check her baby's bathwater. People with MS develop lesions along their spinal cord that affect their mobility and their skin sensations. Lesions also affect the pathways of their autonomic nervous system. Gwen's illness has affected her bowel and bladder functions as well as her sexual responsiveness: the sacral triumvirate.

"I can't feel inside," she explains. "I can't tell that I'm being penetrated I guess is what you'd say. And sometimes I can't feel stimulation on my clitoris."

Yet only six minutes had passed when she pressed the intercom button. The power of vibration to trigger orgasmic reflexes is a mystery and, as we have seen in chapter 10, an occasional boon. Sometimes you don't even have to use it on the usual location. People with spinal cord injuries may develop a compensatory erogenous zone above the level of their injury. (Researchers call it "the hypersensitive area"—or, infrequently, "the oversensitive area.") Applying a vibrator to these spots can have dramatic effects, as documented by Sipski, Barry Komisaruk, and Beverly Whipple. "My whole body feels like it's in my vagina," said the subject, a woman living with quadriplegia who had just had an orgasm—evinced by changes in blood pressure and heart rate—while applying a vibrator to her neck and chest. Komisaruk and Whipple's book *The Science of Orgasm* includes a description of a "knee orgasm" experienced by a young (non-disabled) man with a vibrator pressed to his leg. "The quadriceps muscle of the thigh increased in tension. . . . At the reported orgasmic moment, the leg gave an extensor kick . . . and a forceful grunt was emitted." (In the interest of full disclosure, the young man was stoned.)

I ask Gwen how she made the decision to be part of Sipski's study. "When I first heard about it from my neurologist," she begins, "I thought, Yes, I want to do this. And then I started thinking what the situation was going to be like. And I thought, Well, I don't know if I want to or not. But me and my husband talked it over, and we thought y'all could probably help me." Gwen gets to take a vibrator home with her. The study for which she is a subject includes a treatment component comparing the two stars of the last chapter: the FertiCare (modified with a Woog head) and the Eros. The hope is that vibration (or suction/vibration) therapy can help retrain the sacral reflex arc so that women with spinal issues can reach orgasm more easily.

Gwen retrieves her purse. She asks if we have any other questions for her.

I have one. "Did you hear a loud crash while you were in there?"

"Uh-huh. And talking."

"Sorry about that."

Sipski and I are eating at a suburban Birmingham restaurant where couples drink wine at lunch and seem to have nothing to say to each other. Or maybe they're eavesdropping. I would be.

The lunch conversation has drifted to the topic of non-genital orgasms. The ones that wake you up from dreams.

The ones some epileptics* experience just before a seizure (and that occasionally motivate them to go off their meds). The "thought-orgasms" that ten women had in Beverly Whipple and Barry Komisaruk's Rutgers lab. The individuals Alfred Kinsey interviewed who "have been brought to orgasm by having their eyebrows stroked, or by having the hairs on some other part of their bodies gently blown, or by having pressure applied on the teeth alone." Though in the Kinsey cases, presumably other body parts had been stroked or blown just prior, and the eyebrow and tooth ministerings merely, as Kinsey put it, "provided the additional impetus which is necessary to carry the individual on to orgasm."

I brought along a copy of a letter to the editors of the *British Journal of Psychiatry* entitled "Spontaneous Orgasms—Any Explanation?" The author was inquiring on behalf of a patient, a widowed forty-five-year-old Saudi mother of three, who had "complained bitterly of repeated uncontrolled orgasms." They happened anywhere, at any time, up to thirty times a day, "without any sort of sexual contact." Her social life had been ruined, and she had,

---

*The most interesting being the woman in Taiwan who, once or twice a week, would have an orgasm (followed by a mild nonconvulsive seizure) when she brushed her teeth. The smell of toothpaste alone wouldn't trigger it, nor was it limited to any specific brand. It didn't happen when she poked at her gums with a chopstick or when she moved her empty fist back and forth in a tooth-brushing motion. Curious neurologists at Chang Gung Memorial Hospital gave her a toothbrush and toothpaste and hooked her up to an EEG. Sure enough, after thirty-eight seconds of the "highly specific somatosensory stimulus" we call toothbrushing, it happened. The woman, whose case report appears in a 2003 issue of *Seizure*, was neither delighted nor amused by the situation. She believed she was possessed by demons, and soon switched to mouthwash for her oral hygiene.

understandably, "stopped practicing her regular religious rituals and visiting the holy shrines."

When I look up from the page, the waiter is standing with my gumbo, waiting for me to move my papers. Earlier he came over with the iced teas while Sipski was describing the bulbocavernosus reflex, which tells you whether the sacral reflex arc is intact. The test entails slipping a finger into the patient's rectum and using the other hand to either squeeze the end of the penis or touch the clitoris. If the rectum finger gets squeezed, the reflex is working. The waiters are different in Birmingham than they are in San Francisco, where I eat out. This one said simply, "Who had the unsweetened?"

Sipski's explanation for nongenital orgasms is this: You are triggering the same reflex, just doing it via different pathways. "There's no reason why the impulses couldn't travel down from the brain, rather than up from the genitals." The input would be neurophysiological in the case of epilepsy patients and the Saudi woman, psychological in the case of the Kinsey folks.

Sexual arousal, not just orgasm, reflects this bidirectional split. Here again, spinal cord injuries have helped researchers tease apart the two systems: There is "reflex arousal" and there is "psychogenic arousal." If you show erotic films to someone with a complete injury high up on the spinal cord, the person may say they find the images arousing, but that psychogenic input will be blocked from traveling down the spine, and thus no lubrication (or erection) will ensue. These people can, however, get erections or lubrication from physical, or "reflex," stimulation of their genitals.

Very low spinal cord injuries create the opposite dichotomy: the person can only become lubricated from seeing (or reading or listening to) something erotic. Physical "reflex" arousal is blocked by the injury. Non-disabled men and women respond to both kinds of input (though in women, as we'll see in the next chapter, the head and the genitals are often at odds). Their orgasms can be triggered by a single type of input, or a combination. Barry Komisaruk calls the latter "blended orgasms." This might explain why the single-malt orgasms—vaginal, clitoral, nongenital—all feel somewhat different.

There's one more varietal orgasm I want to ask Sipski about: the kind some kids have climbing the ropes in gym class. Sipski wasn't one of those kids. "I have never heard of this." We both look at each other like we're nuts. I explain that it isn't from contact with the rope, but more from the lifting of your body. Sipski replies that this makes sense, as orgasms from squeezing the pelvic and/or buttock muscles are not unheard of. Kinsey mentions having interviewed some men and "not a few" women who use this technique to arouse themselves and who "may occasionally reach orgasm without the genitalia being touched."

Sipski suspects that this might be how the hands-free orgasm women in the Rutgers lab were managing it. She doesn't know that three weeks before I had lunch with her, I went out for sushi with one of those women. Kim Airs, whose contact information I got from Barry Komisaruk, happened to be in my city visiting friends and agreed to meet to talk about her unique skill set. Airs is a tall, ebullient woman in her forties whose past employers include porn production companies, an escort service, and Harvard University, where she worked with then president Lawrence Summers. Airs learned the "hands-free" technique in 1995, in a breath-and-energy orgasm workshop taught by sex-worker-turned-sex-educator

Annie Sprinkle.* It took her two years to master the craft. Now she can do it easily and upon request, which she does in workshops and talks and, occasionally, on sidewalk benches outside sushi bars.

It was nothing like the *When Harry Met Sally* scene. The people walking past had no idea. She closed her eyes and took some long, slow breaths and after maybe a minute of this, her face flushed pink and she shuddered. If you weren't watching closely, you'd think she was a runner who'd stopped on a bench to catch her wind.

Like the orgasms of Sipski's subjects, those of Airs and Komisaruk's other volunteers were verified by monitoring heart rate and blood pressure. Definitively verifying someone's claim to an orgasm is more difficult than Masters and Johnson would have you believe. The duo described telltale muscle contractions, but Sipski found that not all women have these.† The steep rise and abrupt postorgasm

---

*Highly decorated in both pursuits, Sprinkle holds a Ph.D. in human sexuality as well as a spot on the Adult Star Path of Fame in Edison, New Jersey. The Path of Fame was the brainchild of Mike Drake, manager of the Edison porn emporium Playtime. Drake also oversaw the contents of Playtime's Adult Time Capsule, which include an autographed CyberSkin replica of Sprinkle's vagina. Other items bound to confuse the earthlings of 2069 include nipple clamps, a Decadent Indulgence Vibrator with rotating pleasure beads and "clitoral hummingbird vibrator," and a set of "starter anal beads."

†She figured this out by borrowing a perineometer, a contraction-measuring device first built by Arnold Kegel to document the vaginal strength gains of Kegeling women. To this same end, Kegel made Before and After plaster casts of women's vaginas, to show that their Kegeling regime had rendered them firmer and less "gaping," to use the terminology of Kegel's colleague Marilyn Fithian. "You had to get the plaster out before it got too hard," Fithian told me in an interview years ago. Otherwise, it would get stuck, and no amount of pelvic floor muscle strength was going to help you. "You had to break it inside the vagina," said Fithian, who was in her seventies then and still Kegeling.

drop-off in heart rate and systolic blood pressure are the closest there is to a reliable physiological marker. Airs made the grade.

Sipski is right that at least some of the thought orgasms were helped along by internal muscle flexing. At the end of their paper, Whipple and Komisaruk state that some of the women were making "vigorous muscular movements," and concede that the others may have been doing so more subtly. A how-to Web article under Annie Sprinkle's byline includes directions to squeeze the pelvic floor muscles in order to "stimulate the clitoris and G spot." (Arnold Kegel years ago found that diligent Kegelers tend to have an easier time of orgasm.)

Airs herself, however, described a process involving chakras and waves of energy, but no interior calisthenics. She appeared to be taking herself into an altered state, which makes sense, because that seems to be where people go during an orgasm. Scans show that the brain's higher faculties quiet down, and more primitive structures light up. As in most altered states, people tend to lose their grip on time. In 1985, sex physiologist Roy Levin brought twenty-eight women into his lab and timed their orgasms. After they'd finished, he asked them to estimate how long the orgasm had lasted. With only three exceptions, the estimates were well under the real duration—by an average of thirteen seconds. Orgasm appears to be a state not unlike that of the alien abductees one always hears about, coming to with messy hair and a chunk of time unaccounted for.

What is life like for someone who can discreetly trigger an orgasm with a few moments of mental effort? Airs insists she rarely undertakes it in public. "Sometimes on long plane flights," she said. The last time was while riding the Disneyland tram.

Nor is it, in the privacy of her home, a nightly occurrence. "Usually when I get home I'm too tired."

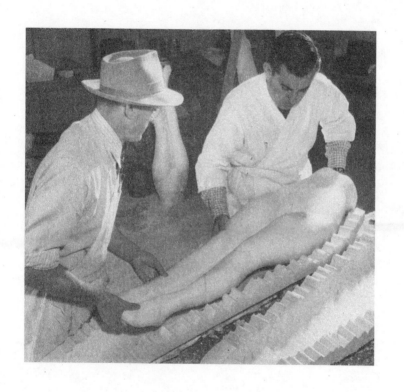

12

Mind over Vagina

Women Are Complicated

t he human vagina is accustomed to visitors. Even the language of anatomy imbues the organ with an innlike hospitality, the entrance to the structure being named the "vaginal vestibule." *Take off your coat and stay awhile.* Gynecologist Robert Latou Dickinson, circa 1910, documented its wondrously accommodating nature, using his fingers as a measuring tool. The volume of the virgin vagina is "one finger"; the married woman rates "two full fingers." Once the babies start coming, it's "three fingers" and up, all the way to Subject No. 163, whose vestibule (and parlor) appear in a pen-and-ink rendering in Dickinson's *Atlas of Human Sex Anatomy* with the doctor's entire hand submerged.

There is no reason why a visit from the acrylic probe of a vaginal photoplethysmograph should be cause for alarm. It's small. It has no sharp edges. It doesn't *do* anything in there except beam a ray of light onto the walls of the vagina. (It does this to measure arousal. The amount reflected back tells you how much blood is in the capillaries; the more light reflected, the more arousal.) There is no reason to say no to an invitation to participate in a photoplethysmograph study at the Female Sexual Psychophysiology Laboratory. And so I have said yes. (Observing someone else who said yes was not an option, because of human subjects review board rules.)

The Female Sexual Psychophysiology Lab is part of the psychology department at the University of Texas at Austin. Its goal is simple but complicated: to untangle the complex, quixotic interplay of body and mind as they pertain to female sexuality. You have no idea what a perplexing mess is female arousal. When a woman is turned on by something or someone, her brain sends a signal to open up more of the capillaries in her womanly recesses. This ups the amount of blood in her vaginal walls, and some of the clear portion of it seeps through the capillaries and coats the vagina. Hello, lubrication. This much we know. But just because a woman is a little moist, that doesn't mean she's going to report feeling aroused. Unlike a man. If a man has an erection, or even half of one, as part of a sex study, he will almost always report that he's aroused. Partly, this is because a boner is easier for its owner to detect than is a damp vestibule. It may also have to do with men's greater skill at detecting subtle physical changes. A 1992 study showed that men were more accurate than women at picking up changes in their heart rate and blood pressure.

Conversely, when a woman *isn't* damp, it needn't mean she's unaroused. To quote Dickinson on the topic of vaginal lubrication,* "Unwise stress has been laid on mucous flow as an adequate gauge of . . . readiness for the entry of

---

*Dickinson's descriptions of the female secretions read, in places, like a WD-40 advert: "It is clear as glass, tenacious and persistent, without being sticky. No other lubricant can compare with it in efficiency for a certain smooth and slippery quality. . . ."

the male. . . . For it must not be overlooked that there are women of strong passion, capable of vigorous orgasm, who show little or no mucous flow." In other words, there can be a puzzling disconnect between mind and body.

In women, the correlation between photoplethysmographic measures of genital engorgement and their own assessment of how aroused they are is so low that some researchers have questioned whether the physical changes alone can be taken to mean anything at all about a woman's state of arousal. Female sexual arousal disorder is rarely diagnosed by photoplethysmograph; it's a conclusion typically reached solely on the basis of a woman having voiced a complaint. The equipment is used mainly for research.

The Female Sexual Psychophysiological Laboratory is run by Cindy Meston. For the past seventeen years, Meston has made a career out of figuring women out—and figuring out how to help them. Possibly longer, if you count the prepsychology era spent traveling rural Canada as a sewing machine sales rep.★ "I'd come to town and do sewing demonstrations and advice programs on the local radio stations. Ladies would call in and say, 'Well, I'm sewing with Ultrasuede, and my stitches are skipping, what should I do?' And I'd go, 'Well you need a Schmetz size-eight stretch needle with a leather tip. . . .'" Meston tried to tell me the name

---

★If you think there's no link between sex and sewing machines, think again. Early sewing machines were treadle-powered, and medical complaints among seamstresses common. Somehow, the men of Victorian medicine decided that the rhythmic pumping of the treadles was arousing women and leading them down the scarlet path to wanton masturbation—and that this "self-abuse" was the cause of their complaints. Wrote J. Langdon Down in the *British Medical Journal* in 1867, "I was struck with the similarity of some of the effects presented to those which my observations at Earlswood had taught me to connect with habits of masturbation." Earlswood meaning the Royal Earlswood Asylum for Idiots. (Where Down was a physician, not a patient. I think.)

of one of the programs, but was laughing so helplessly that it took a while. Finally, she came up for air: "Sewing and Serging with Cindy."

he Female Sexual Psychophysiology Laboratory is on the third floor of the even more time-consumingly named Sarah M. and Charles E. Seay Building. So enthusiastic is the university about its new structure that at one point during construction they set up a Seay Building Web cam, allowing interested parties to log on twenty-four hours a day and watch, literally, the paint dry.

The combination of the Seays' generosity and Meston's flair for interior decor has yielded a science lab not unlike the lobby of the W Hotel. Participant Room 1 (volunteers here are called participants, not subjects) is a small, serenely lit room with modish carpeting and a purple leather recliner. The walls hold framed prints of Modigliani nudes and a flat-screen TV for viewing the inevitable erotic video clips.

It was not always this nice. In the old lab, the erotic films were shown on an ancient TV set whose volume control was on the fritz. It occasionally happened that while Meston or one of her students was running an arousal study subject, department psychologists would be holding a parent-child session in the office next door. "The TV would just be *blaring*," says Meston. "There'd be all this moaning and panting coming through the wall, and we couldn't turn it down!"

The vaginal photoplethysmograph probe that I will be—holding? containing? wearing?—is, for the moment, hygienically sequestered in a Ziploc bag on a table beside the purple chair. The graduate student who is running today's study explains how to insert the probe and directs my attention to a pretty ceramic bowl that in another lifetime might have held sugar packets but now holds little foil packets of "personal lubricant." She doesn't divulge the specifics of her study, because study subjects are always kept in the dark, lest their expectations taint their results. My assignment is simply to sit still and watch a series of video clips—the first batch neutral, the remainder, erotic—while the plethysmograph beams its light beam and measures my physical response to what I'm viewing. (I later learn that I was a control subject in a study about the impact of anxiety disorders upon sexual responsiveness.)

"Okay," says the student. "See you in a bit."

She leaves the room and shuts the door. I take the probe out of the bag. An LED and some wiring are encased in a round-tipped, bullet-shaped piece of clear acrylic. "Cinderella's tampon," I write in my notebook, a notation that I will, weeks later, stare at dumbly for several moments, having no clue what it means. Where the string would be, there is a stiff, plastic-coated cable that leads to a computer. I follow the instructions I was given, and now the cable is curling down the front of my chair. I feel like a bike lock.

On the table is a console similar to the shifter on an automatic transmission. When the video clips begin, I'm to move the indicator up and down according to how I feel. The device is called the arousometer, and Cindy Meston invented it. When the Viagra data started coming in and it appeared that the little blue pill wasn't making women feel more aroused, Meston decided to rethink the methodology. A photoplethysmograph probe collects data sixty times a second. But the psychological data—the woman's own assessment of how aroused she is—was taken only once per subject, at the end of the film footage. "You'd get a questionnaire after the fact, asking, How aroused were you? Did you detect this or this or that? But what, really, is a woman saying when she says, 'I was a four'? Is she telling you what her highest level was? Or did she somehow compute an average of how she was feeling over the whole five minutes?" Meston devised the arousometer as a way for subjects to deliver an ongoing report as the footage rolls.

Using the new device, Meston discovered that there are indeed women who show a nice correlation between body and mind. She is finishing up a study of three groups of women: one with arousal disorder, one with orgasm disorder, and controls. While all three groups' photoplethysmograph responses to the erotic clips were similar, the women with the disorders differed in that they didn't seem to be taking note of the physical changes taking place. And this occurred independent of the level of physical arousal. "Among the [women in the control groups], some had a very small change to the erotic film, but they were detecting it, they were paying attention to it. So with these women, if you could amplify the signal with a drug like Viagra, that could be a good thing. But for a woman who has a normal physical response but she's not attending to it—and really for her the only thing that is arousing is when she feels emotionally bonded or loved or is doing a very specific sexual act—then Viagra is just not going to help." Meston says that although Viagra has not been approved for use by women, doctors often prescribe it anyway—mainly because they don't have anything else to prescribe.

Viagra isn't the only drug being prescribed off-label for women with arousal problems. Los Angeles urologist Jennifer Berman told me some doctors are prescribing low doses of Ritalin. Drugs like Ritalin improve a person's focus, so it stands to reason that it would make it easier to stay attuned to subtle changes taking place in one's body. "It enables a woman to focus on the task at hand," said Berman, managing, though surely not intending, to make sex sound like homework.

In the same category of anecdotal evidence regarding drugs that improve one's focus and make sex more pleasurable and intense, there is pot. As Barry Komisaruk writes in *The Science of Orgasm*, "a substantial proportion of persons claim that marijuana enhances and enriches their sexual experience." For obvious reasons, no one has done a controlled clinical trial of marijuana's effects on arousal and overall sexual satisfaction. Too bad.

The importance of focus fits well with something that Masters and Johnson wrote about back in the 1970s. The team coined the term "spectatoring," which refers to a tendency to observe oneself during sex. Not in an erotic, mirror-on-the-ceiling sort of way, but in a judgmental, critical way. Rather than focusing on the sensations of foreplay and sex, all the feel-good things happening in her body, a spectatoring woman worries about her performance or her appearance. A study by Natalie Dove and Michael Wiederman found that women who were more distracted during sex were—relative to less distracted, more sensation-focused women—less sexually satisfied. They had less consistent orgasms and more often faked them. The questionnaire alone was heartbreaking. Women had to rate how closely a set of statements was characteristic of themselves. Statements such as: "During sexual activity, I worry the whole time that my partner will get turned off by seeing my body without clothes." "While engaged in sexual activity with a partner, I think too much about the way I am moving."

One needn't suffer these particular anxieties to be distracted during sex. A thousand things can play on a woman's mind: work, kids, problems with Ultrasuede. One nonpharmaceutical solution is to teach women to redirect their focus and pay more attention to physical sensations—a practice called mindfulness. A pilot study—meaning it's a preliminary investigation with no control group—by Lori Brotto and two colleagues at the University of British Columbia had promising results. Eighteen women with complaints about their ability to become aroused participated in mindfulness training. Afterward, there was a significant jump in their ratings of how aroused they'd been feeling during sexual encounters.

If it's any solace, even female rats have trouble focusing. I give you a sentence, my favorite sentence in the entire oeuvre of Alfred Kinsey, from *Sexual Behavior in the Human Female*: "Cheese crumbs spread in front of a copulating pair of rats may distract the female, but not the male."

t he largest city of Russia's Pacific coast is Vladivostock, and though it lies in Siberia, the southern location of the city allows it to enjoy ice-free waters year-round. I'm spectatoring the neutral video clips now. There are those ice-free waters. Here's a shot of a statue in the harbor. If the Seay Building Web cam continues to operate, and if one of the cameras is in this room, then someone, somewhere, is very confused. *Looks like she's watching the History Channel with no pants on.*

Abruptly, the neutral footage ends. A snare drum has replaced the oboe and harpsichord music. A man with blonde highlights and a rub-on tan is standing by a desk, dressed in a uniform and a captain's hat. On the desk is a large envelope. The camera, possibly out of habit, zooms in for a full-screen close-up. TOP SECRET, says the envelope. Now here comes more tacky hair and bottled skin, this time of the female variety. It would appear she's going to seduce the captain and steal the envelope.

It is hard to imagine that I will be physically affected

by anything this pair might choose to do to each other. But science has its money on it. A series of studies by Meredith Chivers and colleagues at the Centre for Addiction and Mental Health in Toronto showed that men are more discriminating than women when it comes to how they respond to pornographic images. Women, both lesbian and straight, will show immediate genital arousal (as measured by a photoplethysmograph) in response to films of sexual activity, regardless of who is engaging in it—male, female, gay man, lesbian, or straight. Men, contrary to stereotype, tend to respond in a limited manner; they are aroused only by footage that fits their sexual orientation and interests. (Male arousal is usually measured with a "phallometric" device, which employs a strain gauge to detect changes in the circumference of the penis.) While straight women—and gay men—become physically aroused by footage of two men having sex, straight men generally do not. (A straight man will, however, respond to footage of women having sex, partly because he's looking at two naked women.)

Chivers was struck by what seemed to be "fundamentally different processes" underlying the sexual arousal systems of women and men. To test the limits of the phenomenon, Chivers gamely ran a follow-up study in which men and women viewed, in addition to the usual gamut of human sexual scenarios, footage of bonobos mating.* Here again, the women's genitals responded—though not

---

*You just never know what you're going to see when you sign up for a sexual arousal study. At a sex conference I attended last month, a researcher gave a presentation about an arousal assessment technique called thermography. To make sure that nonsexual reactions like laughter weren't causing the increase in temperature, the subjects' genitals were also thermographically filmed while they watched clips from a Mr. Bean movie.

as strongly as they did to images of human beings—and the men's did not.

And it happens remarkably fast. "Automatic," is how one researcher put it. Or, in Masters and Johnson's cringe-worthy telling, "In a matter of seconds the sexually responding woman may develop sufficient lubrication for coital readiness." The team actually witnessed this happening, using their penis-camera. They describe it in *Human Sexual Response* as "a sweating phenomenon . . . akin to that of a perspiration-beaded forehead." Oddly, the illustration shows a cross-section of a vagina with tear-shaped droplets not beading up on the walls but appearing to rain down into the interior space, as though a summer shower were passing through.*

Women's genitals may respond indiscriminately to images of sex, but the women themselves will often report being totally unaffected by what they've viewed. Based upon how they *feel*, women are quite picky about pornography. Ever since a study by Dutch researcher Ellen Laan showed that women reported significantly higher levels of (subjective) arousal during women-centered porn, arousal researchers started going out of their way to use films made specifically for women.† The one I've been watching is an

---

*Artistic representations of vaginal lubrication tend, historically, toward hyperbole. Old Japanese woodcuts show it sloshing forth as though from a garden hose. Dickinson writes in his *Atlas* that men who played the roles of women in classical Greek comedies would be shown during love scenes with "bags of fluid" hanging between their legs.

†Films by the porn-star-turned-director Candida Royalle are a sex-lab favorite. Meston likes the one about the guy who had the meaning of life tattooed on his erect penis. Alas, he can't get it up (and won't just *tell* anyone what it says), and so a cavalcade of existentially inquiring hotties tries to make big the writing tablet.

example. As unappealing as the captain may be, he certainly is attentive. There was a full minute-long close-up of his tongue doing its thing while due south, his index finger moved in and out—albeit in an unerotic, Schmetz-like manner. The actress, for her part, did her best to seem transported, although every few seconds you'd see her half-open her eyes, like someone cheating at hide-and-seek.

True to Chivers's discoveries, the photoplethysmograph readings for the women in Laan's study were essentially the same during both types of film. Her subjects may have preferred the women-centered clips and perceived them as more stimulating, but their bodies told a different tale.

Be all that as it may, it is the mind that speaks a woman's heart, not the vaginal walls. (Chivers is careful to point out that just because "women demonstrated a genital response to a nonhuman sexual stimulus does not suggest women have a latent preference for sex with animals.") Rape offers a plangent illustration of this fact. I learned in a paper by Roy Levin that rape victims occasionally report having responded physically, even though their emotional state was a mixture of fear, anger, and revulsion. This harkens back to the last chapter, and what Marcalee Sipski learned about arousal from studying women with spinal cord injuries. Lubrication from "reflex arousal" (physical stimulation of the genitals) can occur with absolutely no subjective emotional arousal. Levin also points out that fear causes the release of adrenaline, and adrenaline increases blood flow to the genitals. Which, in turn, enhances lubrication (or erection in men).

Regardless of the mechanisms that may or may not explain a rape victim's physical state, a rapist's defense based upon evidence of arousal has, to quote Levin, "no intrinsic validity and should be disregarded."

· · ·

t he Fruit Machine is a fine example of the perils of trying to make sexual conclusions about folks based solely on measurements of their bodily responses. In the 1950s and '60s, in a laboratory in Canada's National Defence Medical Centre, the government commenced a secret project to pinpoint a simple, trustworthy physical indicator of a man's sexual preferences. John Sawatsky, author of *Men in the Shadows: The RCMP Security Service*, says the goal was to rout gay men from the Royal Canadian Mounted Police and other civil service positions—but to do it scientifically, so that no one was dismissed solely on the basis of hearsay. A bowl of poison with a cherry on top.

What they came up with was pupil response—and an elaborate chair-mounted contraption, nicknamed the Fruit Machine, to measure it. Experiments had shown that people's pupils enlarge when they're interested in what they're looking at. The technique had been used in the past by food marketers to test the appeal of different types of packaging. Canada's Security Service had heard about this and decided to apply it to sexual preferences. If a man's pupils widen while he looks at a naked man, they figured, then that must indicate a preference for the male package.

Trouble set in almost immediately. The team couldn't test the thing because the Department of Defence claimed it had no gay men, and because Mounties were reluctant to volunteer as control subjects. Wisely enough, as it turned out: The scientists had forgotten to take into account differences in the brightness of the images on the screen. When the screen got darker, the viewer's pupil would, naturally, enlarge to let more light in—regardless of who or what the image was. Meaning that if an RCMP recruit looked at an image of a dark brown horse, the Fruit Machine operator would have had to assume that the Mountie and his mount were engaging in something other than national security.

Years after the Fruit Machine project had been shelved, someone got the bright idea of measuring changes in the circumference of a man's penis—rather than of his pupils—while he looked at naked people. Even phallometrics, as this technique is sometimes called, is not a reliable indicator of sexual preference. A strongly motivated man—for instance, an accused pedophile—can learn to control his genital response to an image that he finds erotic.

He can also, with a little training, develop a brand-new one. In 1968, researchers S. Rachman and R. J. Hodgson used classical, Pavlovian conditioning to create a fetish for women's boots. Five men were outfitted with phallometric devices. Over and over, the team showed the men images of nude or provocatively dressed women, followed by images of a pair of knee-high, fur-lined boots. Eventually it worked: In three subjects, the boots by themselves plumped the men's penises as much as the images of the women originally had. Two of the men were also aroused by high-heeled black shoes and by "golden sandals," though no conditioning had been done to these images.

No man got an erection from looking at "brown string sandals."

Many years ago, Cindy Meston, one of Rachman's former students, decided to see if it would be possible to replicate the effect in female subjects. (The incidence of fetishism is far higher in men than in women.) This time the object of the fetish was the voice of the chairman of the psychology department at the University of British Columbia, one Tony Phillips. Meston had taken a snippet from a voiceover of an old student orientation film, and played it over and over as the women looked at erotic images: *Welcome to the department of psychology. . . .* Sadly, for Dr. Phillips anyway, the attempt was unsuccessful.

. . .

If you want to unlock the mysteries of female arousal—the kind the females actually notice and appreciate—the brain is probably the place to turn. After the costly failure of Viagra for women, pharmaceutical companies shifted their attention from drugs that affect genital blood flow to drugs that act directly on the brain. The showiest hopeful to date is bremelanotide, nicknamed "the Barbie drug" because it (a) stimulates the cells that make skin tan, (b) suppresses the appetite, and (c) ups libido. Like Viagra, bremelanotide's sexual properties were discovered by accident—in this case, while the drug was being tested as a sunless tanning agent (under the name Melanotan). The tanning application didn't pan out—"blotchy freckling" and "scrotal moles" are complaints posted by the trial subjects on www.melanotan.org—but some of the women in the study reported feeling randier than usual.

Michael Perelman, director of the Human Sexuality Program at New York's Weill Medical College of Cornell University, ran the most recent bremelanotide trial. Twenty-seven postmenopausal volunteers with female sexual arousal disorder went in for treatment: once with a placebo, the other time with a nasal spritz of bremelanotide. Each time, not knowing what they'd been given, they filled out questionnaires—one shortly after the treatment, another a day later. The drug prompted statistically significant increases in the women's perceptions of how aroused they were after receiving the drug, as well as increases in sexual activity and desire in the twenty-four hours that followed. Bremelanotide is expected to be in Phase III clinical trials (the final stage in the FDA approval process) by late 2008.

The other contender trotted out at the most recent meeting of the International Society for the Study of Women's Sexual Health is a central nervous system drug called flibanserin. This one was originally being tested as an antidepressant. Because diminished libido is a common side effect of antidepressants, researchers were keeping an eye on subjects' sexual feelings. They were surprised to find that flibanserin enhanced, rather than dampened, women's libido. As of 2007, flibanserin is also in Phase III trials.

The FDA tends to be cautious with drugs that affect the brain—especially when they're being used for what some in the medical community view as a lifestyle change. Because of this, and because no one yet understands the mechanism by which flibanserin works, approval may prove thorny.

The drugs come and go: pilot studies, high hopes, fanfare, silence. Apomorphine was the star a couple of conferences back; now you barely hear of it. I asked Cindy Meston whatever happened to it. She laughed. "It made you nauseous."

Fig. 1. The underpant worn by the rat.

13

What Would Allah Say?

The Strange, Brave Career of Ahmed Shafik

d r. Ahmed Shafik wears three-piece suits with gold watch fobs and a diamond stick pin in the lapel. His glasses are the thick, black rectangular style of the Nasser era. He owns a Cairo hospital and lives in a mansion with marble walls. He was nominated for a Nobel Prize.* I don't care about any of this. Shafik won my heart by publishing a paper in *European Urology* in which he investigated the effects of polyester on sexual activity. Ahmed Shafik dressed lab rats in polyester pants.

There were seventy-five rats. They wore their pants for one year. Shafik found that over time the ones dressed

---

*Nominations for a Nobel Prize, I found out when I contacted the Nobel Foundation to try to verify Shafik's, remain secret for fifty years. You make the claim, and nobody can prove otherwise until after you're dead. Add one to your résumé today!

in polyester or poly-cotton blend had sex significantly less often than the rats whose slacks were cotton or wool. (Shafik thinks the reason is that polyester sets up troublesome electrostatic fields in and around the genitals. Having seen an illustration of a rat wearing the pants, I would say there's an equal possibility that it's simply harder to get a date when you dress funny.)

Dr. Shafik published five studies on the effects of wearing polyester, and then moved on to something else. If you print out a list of Shafik's journal articles—and you will need a roll of butcher paper, because there are 1,016 so far—it is hard to say what his specialty is. He has wandered through urology, andrology, sexology, proctology. If you ask him what he is, what he writes under "Occupation" on his tax form, he will smile broadly and exclaim, "I am Ahmed Shafik!"

It is a full-time job. Though Shafik, now seventy-three, is retired from teaching, he continues a heavy schedule of surgery and research, the former funding the latter. (His surgical specialty, as best I can gather, is despots with colorectal issues. He says he has worked on Castro's plumbing, though not recently, and that of the late Mobuto Sese Seko.) Self-funding affords Shafik the freedom to indulge his more esoteric interests*—research projects with no obvious practical ramifications or corporate appeal. In this way he is, as his office manager Margot Yehia has pointed out, a holdover from the nineteenth century, when science

---

*For instance, who else would have funded a study of "the passage of flatus at coitus"? Flaturia, as Shafik has musically named it, is distinct from embarrassing vaginal fart sounds caused by air getting trapped behind the penis during sex. In flaturia, intestinal gas "leaks loudly" from the rectum during sex. Blessedly, it is rare: an affliction of women with a weak internal anal sphincter.

was undertaken simply for the sake of understanding the world.

Shafik's work is far-ranging, but it is not random. The common thread that runs through it is reflexes. In the field of sexology alone, Shafik has planted his flag into twenty new reflexes. If you look at sex through the fabulous black spectacles of Ahmed Shafik, you see more than just a couple expressing their love, or perhaps merely their lust, through the actions of their bodies. You see muscles responding reflexively, without the conscious contributions or consent of their owner, in response to physical stimuli that take place during sex. When a penis hits a cervix in a certain way, for instance, this is a stimulus. In response, a woman's adductor muscles reflexively contract, pulling her thighs together and—in what might be a protective mechanism—limiting the depth of the man's thrusts.

Here's another. When the lower third of a woman's vagina widens—as it does during penetration—several reflexes get triggered. The vaginocavernosus reflex may sound dry or arcane on paper, but it is the basis of what appears to be a remarkable physical synergy between male and female anatomy during sex. When the cavernosus muscles reflexively contract—as they do upon entry—this boosts blood flow to the clitoris. The effect was documented in 1995 by a French team who took Doppler ultrasound images of clitoral blood flow while an inflatable probe was inserted into ten volunteer vaginas.* At the same time as the vaginocavernosus reflex is affecting the clitoris, Shafik found, it's also putting the squeeze on the man's

*While reflexes like the vaginocavernosus may serve to heighten a woman's passion, they cannot stand in place of it. Eight of the ten Frenchwomen "were indifferent" to the overtures of the ballooning pressure probe.

dorsal vein, helping trap blood in the penis and keeping it firm. If there's an intelligent designer in the cosmos, he's got at least one of his priorities straight.

Shafik has published papers on a total of eighty-two ana-tomical reflexes that he has discovered and named. Because other physiologists rarely try to replicate his findings, the reflexive response of the sex research community is to be mildly skeptical. Says Roy Levin, "That man's got more reflexes than I've had hot dinners!" Though Levin concedes that, in general, the study of sexual reflexes has been illumi-nating and worthwhile—at the very least for having "drawn attention to the female reproductive tract as not simply a passive conduit . . . but as a responsive, active canal."

Since each stimulus prompts unique reflexive responses, each must be studied independent of the rest. To mimic an erect penis expanding the opening of the vagina, for instance, Shafik puts a condom-shaped balloon at the end of a catheter, inserts it, and inflates it. Mock bumping of the cervix is done with a sponge on a rod, the sponge having been carved to resemble the head of a penis. (The reflexive responses to these motions are identified via needle elec-trodes in the muscles of the vagina, cervix, uterus, what have you.)

It is noteworthy that Shafik has managed to find doz-ens of women in a Muslim country who will agree to be, say, penetrated by a balloon penis. How does he manage?

I'll know soon. Though no relevant studies are planned for this year, Shafik has arranged for me to see a demonstra-tion of the vaginal reflexes of intercourse. How this will work and on whom they'll demonstrate remains unclear.

On my first morning in Cairo, I wander into a museum near my hotel: the Agriculture Museum. I am the only tourist, a lone adult pushing upstream through currents of happy, shrieking schoolchildren. The museum must have been built around the 1930s and remains charmingly unspoiled by modern advances in museum design. Insect specimens are presented not in their natural environmental niche—e.g., boll weevils on a cotton plant—but in anthropomorphic slice-of-life tableaus: "The Mole Cricket at Home." "The Earwig as a Mother."* The staff taxidermist must have quit at some point, or lost his mind, for some of the animal skins appear not stuffed, but inflated. A sort of hyena pool float hangs on the wall along the staircase, torso bloated, legs sticking straight out from its sides.

I go downstairs to the main exhibit hall, with its full-scale scenes of Egyptian village life: plaster mannequins of men in djellabas, sifting grains and guiding plows. A museum attendant falls into step beside me. He speaks no English, but it's clear he has something to show me. He points to a low wooden door behind a diorama of dusty date-sellers and gestures for me to follow him. He unlocks the door and switches on the lights. We are alone inside a narrow orange-walled hallway that appears to have been, at one time, part of the museum. More village scenes line the sides. Here are women weaving, women telling fortunes, women combing their children's hair. Then I realize: As in real life, the women have been sequestered from men's gazes.

If even inanimate Egyptian women are protected and concealed, how on earth has Ahmed Shafik convinced

---

*The female earwig is renowned for her maternal fastidiousness. She cleans her eggs obsessively with her saliva, which contains an antifungal. If someone—and it is unclear to me who this might be—enters her den and scatters her eggs, she will dutifully gather and repile them. However, if this happens once too often, she will eat them. Even earwigs have their limits.

dozens of flesh-and-blood women to lift up their robes for science?

I pay the guide his baksheesh and go home to take a nap. Around two, I set out again, on foot, to find the Ahmed Shafik Hospital, which I know to be close by. I assumed I could simply ask someone to point me in the right direction, in the same way you can ask anyone in Rochester, Minnesota, how to get to the Mayo Clinic. But hospitals in Cairo are a neighborhood affair, owned by families and small affiliations of doctors and indistinguishable (to the non-Arabic reader) from apartment buildings. I am quickly, deeply lost.

Thankfully, the pay phone (Ringo brand) has not disappeared from Cairo, and with the help of Shafik's faithful factotum Margot, I arrive on time for my first meeting with the man who dressed lab rats in leisure suiting.

A first encounter with Ahmed Shafik is a joyous experience. I am seated on a sofa in his office when he appears in the doorway. He stops in his tracks and stretches out his arms as though in benediction. "Welcome! *Welcome* to Cairo!" Then he steps up and shakes my hand. It's a grand, swinging handshake that begins, like a golf swing, up by his shoulder and finishes in a decisive smacking of palms.

The reflex demonstration is scheduled for the following day, and so we drink coffee and chat. I ask him how he is able to do the sort of work he does in an Islamic country. "First of all," he begins, "I don't publish here. I publish outside. Especially nowadays. In all Arab countries, I don't know why and how, conservative people are coming up greatly. *Greatly!*" He is referring to the recent electoral sweep by the Muslim Brotherhood.

Shafik gains access to women more or less as one does in the Agriculture Museum: baksheesh. The women are sex workers, and he pays them to participate. He pays them in cash and in free medical care—for them and for family members.

"I know a lady, and she helps me. But it is with effort." The research is done in "special flats," where there are also gambling tables. "I go at one or two in the morning. I work the whole night." Not without risk. Sex work is illegal in Egypt. The Ministry of the Interior is sufficiently worked up about it as to have an entire Department for the Prevention of Prostitution. While there is nothing on the books about the legality of paying a woman to let you penetrate her with a balloon, it can't be a simple or pleasant thing to have to explain to an agent of the DPP at two in the morning.

Shafik agrees to put me in touch, via email, with one of his subjects. I contact her several weeks after I get home, with Shafik serving as go-between, sending my questions to her and her answers back to me. The woman refers to the "special flat" as the Home for Prostitutes. The name, as well as the ages of Shafik's subjects (most are in their late thirties), makes it sound like a sort of retirement home for the trade. It's not. Sex workers in Egypt are older than they are in the States; many are middle- and even upper-class women who have been divorced and left with no child support. Raised in an era when women received no education, they turn to sex work as one of the few options to keep themselves afloat and fund their children's educations.

This is not, however, the case with the divorced woman with whom I've been emailing. She does it "simply for having sex." It had never occurred to me that under a religion that forbids sex outside of marriage, prostitution might attract the occasional widow or divorcée. But this is not the reason she lay down with Dr. Shafik's condom balloon. She says she had seen Dr. Shafik on TV and felt that her participation might help women in some way: "I felt very happy when I thought of my participation as my little achievement for science. The peak was when Prof. Shafik showed me the results of the experiments printed in a journal." She clearly holds Shafik in high regard—referring to him as a "world famous surgeon and scientist" and "world-wide well-known Egyptian doctor and researcher"—so much so that at one point I began to picture him, and not her, sitting at a keyboard tapping out the replies.

When I ask this woman to describe the experience, she writes that she was "pretty scared with the sight of the electric apparatuses . . . and with the idea of the needles that were to be inserted into my genital organs and the balloons that were to be placed and inflated in the vagina." As for the test itself, she says simply: "I was not comfortable."

Both religious prohibitions and the law force medical researchers in Muslim countries to take extreme measures. "Even more difficult," says Shafik, "is when you want to do research on a cadaver." Shafik uses a French pronunciation, the accent on the first syllable—CAD-averre. From his mouth, the word sounds foreign and vaguely classy, like a name made up for a car. The Toyota *Cadaverre*. Because of Islamic edicts, there is no tradition of body donation in Muslim countries. Occasionally, Shafik lays hands on an unclaimed corpse, that of a person who has died with no known kin, but more often he has bribed graveyard employees. He is careful to point out that he puts the body back in his trunk and returns it for burial when he is finished.

The conversation trails off, and in that moment I have a realization. I realize that Dr. Shafik's shiny, luxe, peacock-blue suit trousers are synthetic. I can't help myself. I lean forward and pinch a pant leg between my fingers. "Polyester!"

"Yes, yes." Shafik admits it. I am hooting unprofessionally. "Yes, but I tell you . . ." He raises his index finger. "I tell you! Inside is *not* polyester! Underwear, *never!*"

I tell you, there is more functioning technology in one Ahmed Shafik study than in all of Cairo. The ATM machines spit out my bank card like it's gristle. Phone calls from my hotel room must be placed by the desk clerk, who copies down the number and then puts through the call as though I were Claude Rains in *Casablanca*, arranging night passage. The one pedestrian crosswalk I saw in Cairo features a perpetually blinking green man, whom you glimpse in the synapses between speeding cars.

As I walk to my appointment with Dr. Shafik the following afternoon, I try to imagine the scene at the Home for Prostitutes. It is difficult to find a place in this scene for Dr. Shafik and his 12F condom-ended catheter.

We are meeting at his office. The mood is oddly subdued when I arrive. "Mary, I am sorry," says Dr. Shafik in the tone he must use to tell families when operations have not gone as planned. "I asked the house where I go." *Asked* is rendered in two syllables: ASK-ed. This seems to be a regional treatment of the English *k* sound. *Sphinx* comes out SPHINK-us. "To bring you there. I called them last night. They refused! Even the prostitutes, they are very afraid nowadays. I tell you, the religious people are rising up. Up and up! Sex, now, in this country is very secret. The women and the vagina—it's something very criminal."

I got a whiff of this yesterday. A crew from Cairo's English-language TV station came to film a segment about Dr. Shafik. They interviewed me about why I had come to see him but cautioned me not to use the word *sex*. "Say 'sexual intercourse,'" the reporter advised. "Make it sound scientific."

Instead of going to the Home for Prostitutes, we are going one floor down, where someone on the hospital staff has apparently agreed to be a demonstration subject. Dr. Shafik has me wait in the corridor outside an empty ward. Behind the door, voices volley in agitated Arabic. The discussion stops, and Dr. Shafik opens the door. A woman in blue surgical scrubs stands in the corner with her arms crossed.

Dr. Shafik takes me aside. "I am very sorry, but our patient for the reflexes of the vagina . . . She refused!" I am at once dismayed and relieved. No one should have to endure balloon catheters on my account. In place of the woman in blue, Dr. Shafik has recruited a young man, also dressed in scrubs. The man sits on the edge of a hospital bed, looking bored. I cast my mind to the teetering pile of Ahmed Shafik sexology papers on the desk in my hotel room and try to recall which ones pertain to men.

I have a fond hope that Dr. Shafik does not plan to demonstrate the penomotor reflex. When the tip of a penis is stimulated—by bumping against a cervix, say, or the opening to a vagina (or any other orifice, for that matter)—several muscles contract reflexively. Among them are an anal and a urethral sphincter. The closing of the latter prevents urine from mixing with semen in the urethra during ejaculation. The closing of the two together—let's let Dr. Shafik say it—"prevents leak of urine or stools" during sex. Thanking you kindly, penomotor reflex. In his study, Dr. Shafik used a "steel rod . . . covered with a sponge" to stimulate a subject's glans. At the moment, he is holding a telescoping silver pointer. As queasy as this prospect makes me, it would be less awkward than a demonstration of either of the two ejaculation-related reflexes Shafik has published papers on.

One of Shafik's best-known contributions to ejaculatory knowledge was an extraordinary experiment undertaken in 1998 to help determine what it is, precisely, that triggers it. One theory held that ejaculation takes place when the buildup of semen in the prostatic sector of the urethra pushes against its walls with a requisite amount of pressure. (This preorgasm buildup of semen, a sort of massing of the troops from testes, seminal vesicles, and prostate, is called emission; it is emission that creates the sensation of can't-stop-now "ejaculatory inevitability.") Shafik's study cast heavy doubt on the pressure-trigger theory. He inserted a tiny, expandable bulb into his subjects' urethras and found that an expansion of the urethra comparable to what typically happens during emission failed to trigger the telltale muscular contractions of ejaculation. (Roy Levin's guess is that the trigger for ejaculation is the moment when "the summation of all the positive arousing stimuli becomes greater than the negative inhibitor ones.")

Happily, Shafik has in mind something nonejaculatory, something called the cremasteric reflex. He explains how the cremaster muscle automatically raises and lowers the testicles to cool or warm them, depending on the temperature. (The ideal for developing sperm is 95° F.) Shafik did not discover this reflex. It is well known and not the sort of thing one flies all the way to Cairo for. My guess is that he is showing it to me simply to have something to show me.

Shafik addresses the man on the bed, who stands and pulls down his pants. He holds his shirt up out of the way, his hand held flat against his torso. His head is turned to the side, and he gazes stoically into the distance. Despite the circumstances, there is something noble, almost Napoleonic, about his pose. The demonstration is over in a moment, and the man leaves the room. Later, in the lobby café, he will pass by my table and we will pretend not to recognize each other.

Sexually, Egypt today sounds like the United States in the forties and fifties. After the demonstration, I spent some time talking with Saffa El-Kholy, the Egyptian journalist who had come to interview Shafik. The previous year, she told me, she had produced a four-part series on sex that included an invitation to email the program with questions. Although the narration had made it clear that questions would be used anonymously, viewers would often open up a Hotmail account just to pose their questions. El-Kholy heard from women who've "had two orgasms in eight years and aren't sure what the fuss is about." Men who blame their impotence on their wives or, worse, try to keep their wives from having orgasms, so that in the event they (the men) ever become impotent, the wife won't care. El-Kholy: "If you never eat a kiwi, you never want a kiwi."

Though Shafik's research is written up for academic journals, he is comfortable speaking in layman's terms— and does so often for TV. I asked him whether people who hear about his work shun him or find him strange or immoral. "Yes, yes, of course," he replied. "It doesn't discourage me. It's a challenge."

Shafik is similarly untroubled by his low profile in the global community of sex research. Several researchers that I spoke with had not heard of him. In part, this has come about because Shafik does not attend sex-research conferences. And because he only intermittently responds to email. "He's not a team player," says Roy Levin, who long ago gave up attempts at correspondence. The exchange of ideas and the constructive critiques that lie at the heart of Western science make Shafik antsy. He satisfies his own curiosity on a given subject, and then he moves on. As he puts it: "I always never want to go back."

Though Shafik's isolation may compromise his science—or at the very least, his international standing—he is to be commended. As one of the few people in Egypt talking publicly about sex, Shafik performs an even more important role than that of the rogue scientist. If no one on Egyptian TV talks about sex, then no one will talk about it in the café or the bedroom or the doctor's office. Misunderstanding and ignorance will spread. If five hundred unsatisfied women watch Dr. Shafik on television, maybe ten will be encouraged to talk to their husbands. And maybe one or two will eat more kiwis.

14

Monkey Do

The Secret Sway of Hormones

homo sapiens is one of the few species on earth that care if they're seen having sex. The impala is unconcerned. The dingo roundly flaunts it. A masturbating chimpanzee will stare straight at you. To any creature other than you and I and 6 billion other privacy-needing *H. sapiens*, sex is like peeling a mango or scratching your ear. It's just something you do sometimes. This morning, sitting on an observation platform high above a playground-sized rhesus enclosure at the Yerkes National Primate Research Center with researcher Kim Wallen, I have watched a half-dozen monkey couplings, and I'm fairly certain that the situation has caused me more discomfort than it has them.

Wallen and I are here not because of the differences between the sex lives of humans and monkeys, but because of some surprising similarities. Wallen, whom we met in chapter 3, is a professor of behavioral neuroendocrinology at Emory University. He studies sexual desire and the hormones that influence it. Wallen's Female Sexuality Project involves testing different combinations of hormones to see how they effect libido. The hormones are being given to rhesus monkeys, not because monkeys complain about dampened libidos, but because women do—and because monkeys and women have the same hormones, and these hormones affect them in many of the same sorts of ways.

An independent woman may believe herself to be subject to no one and nothing beyond her own volition. And much of the time she is. But there are times, times when certain hormones peak and fertility is at its maximum, that she may find herself behaving in ways that later puzzle her. Hormones can act as the invisible puppet strings behind the discomfiting one-night stand, the shameless flirtation with the bellboy, the unexpected and regrettable kiss between friends. Your genes want you to get pregnant, and hormones are their magic wand.

A dozen studies bear this out. One team of researchers, Stanislaw and Rice, asked 4,000 women to write down the first day within their menstrual cycle when they noticed an increase in sexual desire; the dates peaked at mid-cycle. Women who are part of couples will initiate sex more often at mid-cycle than during the rest of the month—provided they're using a reliable birth-control method (and don't wish to become pregnant); if they're not, then they typically avoid mid-cycle sex. Women also masturbate significantly more often around ovulation than at other times. Take the hormones away, as menopause does, and these mid-cycle spikes in libido level out.

Monkeys offer an unadulterated demonstration of the power of hormones, as the females are not concerned about pregnancy or what their friends will think. Monkeys don't wait until the weekend, or until they've lost two more pounds, or until their roommate is out of town. Here in the rhesus compound, it is much more the case that hormones determine who has sex and when. Their hold over a female animal can be impressive. When they are not close to ovulation, female rhesus monkeys have little to do with males. For the most part, they avoid them. But when they are fertile, they pursue the males constantly, initiating about 80 percent of the sexual encounters they will have.

Right now, in the enclosure below us, the puppet master has control of a shy, skinny monkey whom the researchers have named Page. Page is in heat for the first time. Since we got here, she's been hanging around Keystone, the troupe's boastful, burly alpha male. Alpha males are easy to spot. They are larger by half than the females, and their tails stand in the air like a lion tamer's whip. Lest you forget that Keystone is the alpha male, he does a conspicuous display every ten minutes or so, to remind you of it. He may bounce straight up and down, basketball-like, five or six times. Or he may leap up onto the chain-link fence and shake it by the lapels. It is the rhesus monkey equivalent of karate or doing donuts in the parking lot.

Kim Wallen is not an alpha male sort of guy. He has been married for twenty-five years. He says "crapola" when he misses a freeway exit. Today he is dressed in chinos and hiking shoes. His shirt is quietly checkered, and his tie has two small spots on it from the fish soup he ate for lunch. Perhaps because of the photo of him on the Emory Web site, in which he is leaning against a large tree trunk, a woman once wrote to him, "You look like a man who'd like to go for a walk."

A hawk circles above us, and the monkeys hoot and clamor. It's the sound that the crowd makes when George Clooney arrives on the red carpet outside the Oscars. We watch as Page gradually, subtly, moves closer to Keystone.

"They do what looks like a random walk, but each time they stop, they're a little closer. It's like the teenage dance, where you're interested in a guy and you kind of hang around in the area, waiting to go get punch until he goes up to get punch." Wallen says he and his colleagues sometimes entertain themselves by going to bars and trying to guess who'll end up with whom at the end of the evening, based on their behavior early in the evening. "It's exactly like following these animals." Page is now four or five feet from Keystone, picking up a rock, as though that rock were the reason she crossed the enclosure just now.

Why the coyness and hesitation? In Page's case, it has to do with her low rank and the risks that go along with it. If she's too obvious in her solicitations, she stands a chance of being thwacked by a higher-ranking female. Furthermore, adult male rhesus monkeys—if you're a female rhesus—are big and intimidating. "Imagine it," says Wallen. "You're this little teeny female, you've done nothing with the adult males for all of your preadolescent period, and all of a sudden you wake up one day and say, 'You know, this guy is really attractive.'"

Page just sat down a foot away from Keystone, at the top of a climbing structure. In the world of monkeys, this counts as a come-on. Primate researchers call it "initiating proximity." I'm actually feeling nervous for her. Wallen leans forward in his chair. "If you watch, sometimes you can actually see their hands shake."

A higher-ranked female named Gawk just lunged at Page. "If Page didn't have strong enough motivation"—i.e., the push provided by hormones—"it—sex—would just never happen," Wallen says. If you take away the complicated, anxiety-provoking social structure that exists in the Yerkes compound (or in the wild), hormones cease to matter as much. A lone male and a lone female monkey placed in an enclosure together will get down to business in no time.* There's no risk involved.

Among women too, socially risky sex tends to happen when the hormones hit their peak. In a study by M. A. Bellis and R. R. Baker, the sex that cheating women were having with their lovers closely mirrored their monthly cycle, peaking on the day of maximum fertility. But the sex these women were having with their husbands was randomly distributed throughout the month. The women's hormones, it appears, were providing the extra impetus needed to take on the risk of getting caught.

"Look!" I shout to Wallen, who has turned away to speak with a graduate student who has joined us, inputting behavior observations on a laptop. Keystone has stepped up onto a female's back legs as though they were stilts. "They're doing it!"

"That's not Page," says Wallen. "That's Tequila." Tequila is the beta female. As a high-ranked female, she gets mounted out of courtesy. Meanwhile, Page is parked at the feeding ledge, seemingly consoling herself with Monkey

---

*Ditto humans. In 1973, researchers put a group of students (who'd never met) in a pitch-dark room for an hour, after telling them each would leave alone and never see the others. In other words, no judgment, no consequences. Meanwhile, infrared cameras were rolling. Ninety percent touched a stranger, 50 percent hugged one, and an unspecified number "necked." When the experiment was repeated in a lighted room, no one made physical contact.

A boy who admitted to necking with a stranger named Beth said: "We expressed it as showing 'love' to each other. Before I was taken out, we decided to pass our 'love' on. So . . . Laurie took her place."

"People share strong yearnings to be close to each other," concluded the authors. "However, social norms make it too costly to express these feelings. Perhaps these traditions have outlived their usefulness." Oh, probably not.

Chow. One eye remains on the Keystone scenario. As soon as Tequila moves away, Page sets out again. She stops near Keystone. She slides her hand toward him.

In the social lexicon of the rhesus, this is a subtle come-on, or "present" (short for "presentation").* Less subtle overtures include moving one's tail out of the way, touching the male, or gently slapping the ground in front of him. Part of the reason it took primatologists—who were, in pre-Jane Goodall days, all men—so long to acknowledge the female rhesus monkey's role in initiating sex was that the solicitations were so, well, forward. "There was a very strong predisposition not to be looking for that," Wallen says. The pioneering primatologist C. R. Carpenter first documented the hand slap as a female sex solicitation in the 1940s, but his papers were ignored for years.

I look down at my notepad and when I look up again, I catch Keystone and Page in the act. Despite the protracted buildup, it's not the least bit arousing to watch. Monkeys have sex the way we pump a keg or fluff a pillow: a brief series of repetitive actions undertaken with no discernible passion or emotion, and not a terrific amount of interest.

Over the next few minutes, Keystone mounts Page repeatedly, but always seems to lose interest after a few desultory thrusts. Wallen explains that rhesus monkeys are "multimount, multi-intromission ejaculators." He'll be on her and off her five or ten times before he finishes up.

"There goes the mount. That's an intromission.

---

*A visit to Yerkes will forever after distort your image of corporate America. On my flight home, the woman behind me was talking about the *presentation* she was planning for a man named Mark. Her seatmate had just finished up a series of *displays* at the regional sales conference.

Three pelvic thrusts. Now the dismount." If Yerkes ever loses its funding, Wallen could find work as an Olympics commentator.

This kind of furtive, piecemeal copulation might have evolved as a way of passing on your genes while at the same time avoiding a possibly life-threatening fight with another male. The sneakier you are about it, the less attention you attract and the less jealousy you provoke, and the longer you live.

The other strategy is to be a speedy ejaculator: in and done before the other males notice what you're up to. Male chimps are tops at this. A research paper on the origins of premature ejaculation (PE) states that chimpanzees ejaculate within an average of seven seconds after they mount a female.* The author speculates that it is perhaps because of this that chimpanzees are known for a lack of aggression among males during mating season. It's hard to get irritable over a liaison that takes less time to finish than a banana.

This author lists the human male's average time lapse between penetration and orgasm as two minutes—placing him midway between the chimp and the orangutan (eleven minutes). U.K. sex physiologist Roy Levin puts two minutes at the fringes of normal; his figure for an average male's thrusting time is two to five minutes (or, if you prefer, 100 to 500 thrusts). In the latest papers on premature ejaculation, two minutes falls a scant half minute outside

---

*Ladies, do *not* get involved with a chimp. Not only are they fast ejaculators, they want to perform this minor irritation *constantly*. (Highest copulatory frequency of all primates.) And here's how they let you know: "the male invitation posture," in which the male sits on the ground, knees up and legs wide open to, quoting *Reproductive Biology of the Great Apes*, "reveal his erect penis." As an alternate wooing strategy, the male chimpanzee will shake a branch at you.

the category "probable premature ejaculation."* Does our author have a personal premature-ejaculation ax to grind? "If premature ejaculation was normal and advantageous in the past, . . . why is it labeled *dysfunction* today? . . ." he writes tellingly. "If rapid ejaculation is normal, then premature ejaculation by itself should not be of clinical concern unless it is extreme, such as occurring before intromission." The author advocates placing more emphasis on "the tender touch, the passionate caress, the gentle rub, the titillating probe"—all of the things men do better than orangutans—rather than making men miserable by asking them to try to expand their ejaculatory latency. A point well taken, but still and all, sympathies to the Mrs.

Sexual desire is a state not unlike hunger. You may find yourself getting up for a snack long before you're aware of a physical sensation. If you are a single woman midway through her cycle, you may find yourself on a barstool or a set of front steps you swore you'd never climb again. In the words of young Page (via the mouthpiece of Kim Wallen), "I don't know what I'm doing here, but here I am."

Wallen is leaning back in his chair, with one foot on a rusted railing. "I have no concept at all of what's attractive in a rhesus," he is saying. "But I have seen males mate

---

*A category defined by PE expert Marcel Waldinger as one to one and a half minutes Intravaginal Ejaculation Latency Time (IELT)—a high-octane way of saying How Long He's in Before He's Done. "Definite" premature ejaculation is defined as consistently less than one minute. (Certain antidepressants are being used these days to treat PE—including some that can be taken a couple hours before sex.)

You never hear much about the opposite condition, delayed ejaculation. Possibly, this is because for years it was called "retarded ejaculation," and who wants to admit they've got that?

with females that struck me as incredibly *un*attractive." And vice versa. This is what primate sex hormones do: "They make individuals perceive other individuals as more attractive than they'd normally perceive them." Hormones are nature's three bottles of beer.

But not if you're on the Pill. In humans, a hormone-based contraceptive levels out the monthly peaks and troughs of one's natural hormone levels—and, in consequence, those of libido. The Pill supplies a steady daily dose of hormones, enough that your body stops supplying its own unsteady, cyclically fluctuating dose. While the Pill's estrogen levels are high enough to prevent ovulation, they are lower than a natural mid-cycle peak. Says urologist and sex advice author Jennifer Berman, "The Pill basically puts you into a kind of menopausal state."

The Pill contains estrogen and progesterone, but it also affects testosterone. And it is testosterone, more than any other hormone, that influences a woman's libido. What the Pill does, specifically, is raise levels of sex-hormone-binding globulin (SHBG), a protein in the blood that binds itself to testosterone, taking the hormone out of commission. And going off the Pill might not restore libido. In a 2006 study, urologist Irwin Goldstein looked at women's levels of SHBG and of their free (unbound) testosterone while they were on the Pill and after they'd gone off it. Their SHBG didn't decrease after they stopped taking the Pill, and their testosterone levels—and, presumably, their libido—didn't recover.

Why hasn't low libido been listed as a side effect for oral contraceptives? "The FDA doesn't consider behavior and in particular sexual behavior to be something they're concerned about," says Wallen. And why don't doctors mention it to women before they pick up the prescription pad? In part because not that many women on the Pill complain about low libido. One in four is the statistic I've heard. For many women, the freedom from worrying about pregnancy cancels out any mid-cycle dip in libido; they're having more sex then, not less. The Pill doesn't make women enjoy sex less, it doesn't change their responsiveness; it just mutes their drive. A lot of them don't even notice, and for some, it's a price worth paying.

Menopause is a natural, more exaggerated version of being on the Pill. Estrogen and testosterone levels fall, taking libido along with them. As a solution to flagging sexual desire in postmenopausal women, Procter & Gamble, in 2004, came up with a testosterone patch, called Intrinsa. When the FDA asked for more safety data, the company dropped their plans—no doubt skittish after the sudden crash-and-burn of postmenopausal hormone-replacement therapy. But in July 2007, the EMEA, Europe's version of the FDA, went ahead and approved Intrinsa for women with hypoactive sexual desire disorder (low libido). Americans who want to try it need look no further than the closest Internet pharmacy.

The testosterone patch was the subject of vigorous debate at the 2007 meeting of the International Society for the Study of Women's Sexual Health (held, to no one's amusement but my own, at Disney World). An endocrinologist's pro-patch presentation was met with vocal concerns that researchers and their pharmaceutical company sponsors were, as one attendee put it, "making a normal midlife phenomenon into a disease."

If only rhesus monkeys could read some of the human studies they've inspired. What would they make, for instance, of sixty-two married American women, for three months running, smearing synthesized rhesus monkey "sex pheromones" onto their chests before getting into bed? The smears did not, it turned out, inspire the husbands to have sex with their wives more often than usual. They did not inspire anything at all, except possibly the idiom "There's something I need to get off my chest."

Why did sixty-two American women do this? The short answer is that the researchers paid them. (One dollar a day. This was 1977.) The long answer is that monkey-observing scientists used to believe that the reason rhesus monkeys have more sex around the time the females ovulate is not that the female is under the sway of hormones that push her to make a move, but rather that the female has pheromones—chemical triggers of behavior—that prompt the males to make a move. (Sex pheromones are commonplace in other neighborhoods of the animal kingdom—among insects, for instance, and rodents and swine—but olfactory sex triggers for primates were until this point unknown.)

The rickety notion of rhesus—and, by implication, human—sex pheromones can be traced to a rhesus monkey research colony in the U.K. and to the behavioral neuroendocrinologist who observed it. In 1971, Richard Michael claimed to have pinpointed compounds in the vaginal secretions of his females that, when sniffed, caused the male monkeys to initiate sex. (But not very many of them. Critics point out that just two males accounted for 50 percent of the data.) Michael called the purported rhesus pheromones "copulins,"* a word I cannot write without

---

*Shortly after his discovery, Michael patented the chemical makeup of the copulins as a sex attractant. George Preti, a researcher at the Monell Chemical Senses Center in Philadelphia, says a major fragrance company in the seventies was said to be adding synthesized monkey copulins to its perfumes. Which seems disgusting, until you learn that another ersatz sex attractant fragrance, called Realm, was made from skin compounds derived from scrapings off of, quoting Preti, "the inside of casts worn by injured skiers."

picturing a race of small, randy beings taken aboard the starship *Enterprise*.

Other endocrinologists had doubts about Michael's claim. D. A. Goldfoot and three colleagues at the Wisconsin Regional Primate Research Center made "lavages"—a pretty French word for the liquid they got from washing out the vaginas of rhesus monkeys in heat—and smeared the substance onto the back ends of neutered (i.e., non-hormone-producing) females. The expectation was that if copulins were for real, the males would try to mate with the lavage-anointed, neutered females. The males did not.

Michael's sex-pheromone work got tremendous media coverage nonetheless, which is unfortunate, as it sent our understanding of female hormones and female sexual behavior way off down the wrong boulevard. It implied that when it came to sex, the female primate was a passive receptacle with no drive or interest of her own.

However, I cannot hold this against Dr. Michael, for his work inspired a highly diverting period of scientific inquiry. In 1975, for example, a team of researchers from the Monell Chemical Senses Center in Philadelphia launched an investigation of changes in the "pleasantness" of women's vaginal odors across their monthly cycle. Seventy-eight subjects were asked to sniff tampons that four women had worn during the various phases of their cycle. (For obvious reasons, the women were asked not to eat onions, garlic, or asparagus for the duration of the study. Less obviously, the women were discouraged from eating broccoli, brussels sprouts, cabbage, chili, curry, kale, sauerkraut, and pineapple.) The supposition was that the odors might be more appealing during a woman's ovulatory phase than at other times during her cycle. And they were: Subjects judged them slightly more pleasant and less intense than at other times. However, the authors reported, the data did not go so far as to "support the notion . . . that vaginal secretion odors are particularly pleasant to human males."*

There was one other nominee for human sex-pheromone status: A compound called androstenone was found to exist in men's underarm sweat. Androstenone had long been known as a potent swine sex pheromone; when a pig in heat sniffs it, she becomes receptive to being mounted by a boar. Hence its presence in male bodily secretions sent endocrinologists into frenzies of speculation. Its actual effect on women proved unclear—though not for want of trying. For years, psychologists and endocrinologists took to sneaking around in public spaces spraying furniture and bathroom stall doors with cans of Boarmate, a synthetic, aerosolized version of androstenone.

Occasionally, the studies seemed to turn up an effect. M. D. Kirk-Smith and a colleague at the University of Birmingham in the U.K. sprayed Boarmate on what had been determined to be an unpopular seat among women visitors to a dentist's waiting room. The aim was to see if more women would now be attracted to this chair. The seat's popularity was secretly observed by receptionists. The Boarmate appeared to work its charms on the women, who sat in the chair significantly more often than they had before Kirk-Smith and the can of Boarmate hit the scene.

---

*Preti and gang were clearly not speaking for the millions of soiled-panty enthusiasts in our midst. A Google search on "soiled panties" produced 78,000 hits, most of them directing you to freelance sellers, women who throw up a Web site with a couple of photos and a PayPal link. Wikipedia says some Japanese sex shops operate panty exchanges for girls, who wear a pair overnight and then exchange them for a new pair on their way to school. "The more soiled they are, the more they will fetch at sale," says Wikipedia, yet further distancing itself from stuffy rival Britannica.

What does the dentist chair project prove about women and men and their interactions with each other? Nothing, says George Preti, of the Monell Chemical Senses Center, who was dismissive of the study and critical of its methodology. The bottom line is that men's armpit secretions are unlikely to serve as an attractant to any species other than the research psychologist.

Despite the washy evidence that androstenone has an effect on human sexual behavior, it wasn't long before someone patented an androstenone-based human sex attractant. Winnifred Cutler used to work with George Preti. The two parted ways when she began placing ads in the backs of men's magazines for Athena Pheromone 10X ("Raise the Octane of Your Aftershave"). Cutler published a study stating that men who added 10X to their cologne were having significantly more dates and more sex than a control group. She concluded that her product had made the men more sexually attractive. Preti, in turn, claimed that Cutler had failed to demonstrate solid evidence that this was so. And to this day, no amount of 10X can bring the two together at industry gatherings.

I have a better suggestion for Cutler's customers. Stop wearing cologne. Women don't find it attractive. If you don't believe me, here is a quote from a press release from the Smell and Taste Treatment and Research Foundation in Chicago: "Men's colognes actually reduced vaginal blood flow." Foundation director Al Hirsch hooked women up to a vaginal photoplethysmograph and had them wear surgical masks scented with ten different aromas or combinations of aromas. (To be sure the women weren't just getting aroused by dressing up in surgical masks, Hirsch put unscented masks onto a control group.) In addition to the smell of cologne, the women were turned off by the scent of cherry and of "charcoal barbeque meat." At the top of the women's turn-on list was, mysteriously, a mixture of cucumber and Good 'n' Plenty candy. It was said to increase vaginal blood flow by 13 percent.

Though the existence of human pheromones remains open to debate, sexuality does seem to play a role in how men and women respond to the scent of each other's hormones. Researcher Ivanka Savic of Stockholm's Karolinska Institute asked straight women and gay men to sniff a particular hormonal component of male sweat. As they did so, their hypothalamus lit up on a PET brain scan, suggesting a sexual response rather than just an olfactory one. The same kind of brain response showed up when Savic had straight men—and, in a second study, lesbians—sniff an estrogen-like compound found in women's urine. Savic emphasized that the sweat and urine compounds did not—as would a true pheromone—prompt any changes in behavior, except, possibly, refraining from signing up for future Ivanka Savic studies.

15

"Persons Studied in Pairs"

The Lab That Uncovered Great Sex

W hen I began this book, I harbored a naïve fantasy that I would find a team of scientists working to discover the secret to amazing, mind-rippling sex. They would report to work late at night in a windowless, hi-tech laboratory and have unplaceable accents and penetrating stares. Week after week, couples would be hooked up to instruments, measured, interviewed, filmed. Data would be analyzed, footage reviewed, and one day one of the researchers would set down her pen and nod knowingly.

I suspected that the secrets uncovered in this lab would have less to do with vasocongestion or vaginoclitoral distance or hormones than with how the two people on the bed in the laboratory felt—about one another, and about sex. And that those feelings would color and inspire the things they did. And that without those feelings you could play the overture and hit the crescendos just fine, but the music would not take you to the same rapturous place.

One day, with only two months to go before I turned in the manuscript, I found that lab. In 1979, William Masters and Virginia Johnson published *Homosexuality in Perspective*, a book to which I had never before seen or heard a reference. For five years, Masters and Johnson observed and compared the laboratory sexual encounters of straight, gay male and lesbian, and "ambisexual" couples. (The team coined the term to refer to nonmonogamous sexual opportunists who show no preference between men and women throughout their very busy sex lives.)

To keep the subjects' identities secret, the researchers did indeed schedule the sessions late at night or on weekends, when no one was in the building. The rest of it surpassed even my own imagination. While some of the subjects were having sex with their spouses or long-term partners, others were doing it with a stranger—not a stranger of their choosing, but one assigned to them by Masters and Johnson. These latter men and women would show up at the lab, chat with the researchers, and, following a short orientation session, get down to business with a man or woman they had never before laid eyes upon. While Masters and Johnson observed.

I learned about the project in a *New York Times* health column. Jane Brody had described the book and its conclusions the week it came out. The subheads the paper had supplied were vague and coy:* "Persons Studied in Pairs,"

---

*Except for this one: "Rape Fantasies for Both." Masters and Johnson published a list of the top five sexual fantasies of the gay and straight men and women. Forced Sexual Encounters was either No. 1 or No. 2 for all four groups. Both straight men and lesbians imagined themselves interchangeably in the role of rapist and "rapee." In the case of gang rape fantasies, gay men occasionally "played an additional role of planner or organizer."

said one. It was like writing up the Million Man March under the headline "Persons Walking in a Group." In a sentence at the end of a paragraph describing study protocols, Brody notes simply: "Some were assigned partners." The casual reader, alighting here, might have mistaken the column for a piece about square dancing. I immediately tracked down a copy of the book.

As always, and like most sex researchers, Masters and Johnson were stingy with the irrelevant details. I can tell you that the thermostat was set at 78, presumably because the couples were naked and, of course, had no covers over them. I can tell you that some of the participants asked for background music, though I cannot tell you which albums, just as I could not tell you the titles of the "stimulative literature" used to arouse subjects in *Human Sexual Response* twenty years before.

The team did mention that many of the men and women who had been assigned a partner worried that this person wouldn't find them attractive. Oddly, the reverse anxiety never surfaced—no one seemed concerned about whether they themselves would feel any attraction to the stranger whose genitals they were about to experience in almost every way imaginable: manually, orally, coitionally.* Catching something wasn't a concern, because everyone was screened for venereal disease, and AIDS hadn't yet surfaced. The researchers themselves had but one qualm. They worried at first that some of the subjects might come on to them

---

*But not rectally. Anal sex—which Masters and Johnson dubbed "rectal coition"—was shunted off to its own little investigation. Twelve couples were observed: five gay male and seven straight (three of the latter being strangers). The researchers studied anal sphincter contractions (conclusion: it mainly just hurts at first) and looked to see if arousal causes rectal lubrication (it doesn't).

and/or make small talk, I cannot tell which, for they phrased it as "the problem of study subjects attempting social interchange" with the researchers.

Unlike *Human Sexual Response*, this project did not primarily concern itself with the physiology of arousal and orgasm. Everything Masters and Johnson had observed in their heterosexual subjects in the fifties (a subset of whom became the later project's hetero group), they found, applied to their gay and lesbian subjects. Having now observed "hundreds of cycles of sexual response" in gay men and lesbians as well, they quickly concluded that arousal and orgasm are arousal and orgasm, whether a couple has one, two, or zero penises between them.

A large chunk of the book is spent comparing "functional efficiency" and "failure incidence" of the different groups: gay male versus straight versus ambi, long-term versus assigned. Table after table with titles like "Functional Efficiency of Ambisexuals in Manipulative Stimulation and Coition." This was Masters and Johnson as their critics saw them: the mechanizers of sex, obsessively focused on "effective stimulation," reducing passion to a series of impersonal physical manipulations.

But ultimately the team set aside their stopwatches and data charts and turned a qualitative eye upon their volunteers. What emerged were two portraits. There was efficient sex—skillful, efficient, goal-directed, uninhibited, and with a very low "failure incidence." Here there were no significant differences among the study groups. Basically, anyone who signed on as a Masters and Johnson volunteer—gay men, lesbians, straight, committed or not—tended to have, as they say, 100 percent orgasmic return. Because really, why would people who knew themselves to be iffy responders volunteer for this project?

But efficient sex was not amazing sex. The best sex

going on in Masters and Johnson's lab was the sex being had by the committed gay male and lesbian couples. Not because they were practicing special secret gay male and lesbian sex techniques, but because they *took their time.* They lost themselves—in each other, and in sex. They "tended to move slowly . . . and to linger at . . . [each] stage of stimulative response, making each step in tension increment something to be appreciated. . . ." They teased each other "in an obvious effort to prolong the stimula-tee's high levels of sexual excitation."

Another difference was that the lesbians were almost as aroused by what they were doing to their partner as was the partner herself. Not just because, say, fondling a breast turned them on, but because their partners' reactions did. Masters and Johnson's heterosexual subjects failed to grasp that if you lost yourself in the tease—in the pleasure and power of turning someone on—that that could be as arousing as being teased and turned on oneself. "Not only were committed lesbians more effective in satisfying their partners, they usu-ally involved themselves without restraint . . . far more than husbands approaching their wives." The straight man, in most cases, "became so involved in his own sexual tensions that he seemed relatively unaware of the degree of his part-ner's sexual involvement. There were only a few instances when the husband seemed fully aware of his wife's levels of sexual excitation and helped her to expand her pleasure . . . rather than attempting to force her rapidly to higher levels of sexual involvement."

The same criticisms applied to straight women: "This sense of goal orientation, of trying to get something done . . . was exhibited almost as frequently by the heterosex-ual women as by their male partners." They ignored their husband's nipples and just about everything else other than his penis. Meanwhile, the gay men lavished attention on their partners' entire bodies. And the gay men, like the lesbians, were adept at the tease. Unlike the wives: "Rarely did a wife identify her husband's preorgasmic stage . . . and suspend him at this high level of sexual excitation. . . ."

Masters points out that the heterosexual couples were at a disadvantage, as they do not benefit from what he called "gender empathy." Doing unto your partner as you would do unto yourself only works well when you're a gay man or lesbian. "Since rapid forceful stroking was the pattern of choice during male masturbation," Masters wrote, "it was also a consistent pattern during the male's manipulation of his female partner's clitoris." The lesbians' lighter touch was "generally the more acceptable. . . ." For no doubt similar reasons, the straight women, their husbands told the researchers, "did not grasp the shaft of the penis tightly enough."

But the empathy gap is not insurmountable. One has only to speak one's mind. The other hugely important difference Masters and Johnson found between the heterosexual and gay male and lesbian couples was that the gay male and lesbian couples talked far more easily, often, and openly about what they did and didn't enjoy. Gay men and lesbians simply seemed more comfortable in the world of sex. Masters gives the example of the heterosexual men's finger insertions: "Though many heterosexual women evidenced little pleasure . . . and were obviously distracted by [it], . . . only twice did they ask their husbands to desist."

It seems to me that heterosexual people have come a long way since 1979. The media's ubiquitous coverage of sex and sex research—as well as the genesis and population explosion of TV, radio, and newspaper sex advisors—have chipped away at the taboos that kept couples from talking openly with each other about the sex they were having. Bit by bit, sex research has unraveled the hows, whys, why-nots, and how-betters of arousal and orgasm. The more the researchers and the sexperts and the reporters talked about sex, the easier it became for everyone else to. As communication eases and knowledge grows, inhibitions dissolve and confidence takes root.

Sadly, the main thing people recall about *Homosexuality in Perspective*, if they recall anything at all, is that Masters and Johnson spent the second half of the book touting a therapy for helping gay men and lesbians convert to a heterosexual orientation. The team went out of their way to assure readers that they screened clients carefully, accepting only those who had changed their sexual orientation to gay male or lesbian after a traumatic experience (rape or abuse, for instance). They insisted that no gay man or lesbian who came to them for therapy was ever pressured or encouraged to pursue a heterosexual orientation. However, as one critic pointed out, many should probably have been encouraged *not* to pursue it.

But let's give Masters and Johnson their due. And while we're at it, Alfred Kinsey and Robert Latou Dickinson and Old Dad and everyone else in these pages. The laboratory study of sex has never been an easy, safe, or well-paid undertaking. Study by study, the gains may seem small and occasionally silly, but the aggregation of all that has been learned, the lurching tango of academe and popular culture, has led us to a happier place. Hats and pants off to you all.

acknowledgments

S ex research is a little like sex in that most people who engage in it are more comfortable without an audience. The researchers who invited me into their labs did so at the peril of their funding, their privacy, their academic standing, their sanity. For saying "yes" when "no" was the sensible answer, I am deeply grateful to Jing Deng, Anne Marie Hedeboe, Geng-Long Hsu, Barry Komisaruk, Roy Levin, Ken Maravilla, Ahmed Shafik, Marcalee Sipski, and Margot Yehia. I am extravagantly indebted to Kim Wallen for his contributions to not one, but two, chapters, and to Cindy Meston, my sex research swami, for all the help, hospitality, and hilarity.

For graciously enduring my demands on their time, I thank Kim Airs, Jennifer Bass, Irwin Goldstein, Stephanie Mann, Robert Nachtigall, Michael Perelman, Anne Pigue, Carol Queen, Harold Reed, Arlene Shaner, Ira Sharlip, Marty Tucker, and Alice Wen.

This is my third book with W. W. Norton, and there is good reason for that. The list of Norton folks to whom I'm indebted is practically their phone directory. Boldface must be applied to a few of those names: Jill Bialosky, whose editing pencil should be bronzed should she ever retire (an event I will do everything in my power to prevent); Erin Lovett and Winfrida Mbewe, who make me feel bad for every author whose book is launched by someone else; and Bill Rusin, a man born to sell books. I thank you all for your commitment, creativity, and enthusiasm.

Beyond the halls of Norton, praises must ring for photo curator Deirdre O'Dwyer. My agent, Jay Mandel, deserves another 15 percent, and my husband, Ed, deserves a medal.

bibliography

Foreplay

Allgeier, Elizabeth Rice. "The Personal Perils of Sex Researchers: Vern Bullough and William Masters." *SIECUS Report,* March 1984, pp. 16–19.

one | The Sausage, the Porcupine, and the Agreeable Mrs. G.

Boas, Ernst P., and Ernst. F. Goldschmidt. *The Heart Rate.* Springfield, Ill.: Charles C. Thomas, 1932.

Buckley, Kerry W. *Mechanical Man: John Broadus Watson and the Beginnings of Behaviorism.* New York: Guilford Press, 1989.

Cohen, David. *J. B. Watson, the Founder of Behaviorism: A Biography.* London, Boston: Routledge & Kegan Paul, 1979.

Dickinson, Robert Latou. *Atlas of Human Sex Anatomy.* Baltimore: Williams & Wilkins, 1949.

Gathorne-Hardy, Jonathan. *Sex: The Measure of All Things.* Bloomington: Indiana University Press, 2000.

Jones, James H. *Alfred C. Kinsey: A Public/Private Life.* New York: W. W. Norton, 1997.

Kinsey, Alfred C., et al. *Sexual Behavior in the Human Female.* Philadelphia: W. B. Saunders, 1953.

Klumbies, G., and H. Kleinsorge. "Circulatory Dangers & Prophylaxis During Orgasm." In *Sex, Society and the Individual.* Bombay: International Journal of Sexology, 1953.

Magoun, H. W. "John B. Watson and the Study of Human Sexual Behavior." *Journal of Sex Research* 17 (4): 368–378 (1981).

Masters, William H., and Virginia E. Johnson. *Human Sexual Response.* Boston: Little, Brown, 1966.

Pomeroy, Wardell. *Dr. Kinsey and the Institute for Sex Research.* New York: Harper & Row, 1972.

two | Dating the Penis-Camera

Alzate, Heli, and Maria Ladi Londoño. "Vaginal Erotic Sensitivity." *Journal of Sex and Marital Therapy* 10 (1): 49–57 (1984).

Archibald, Timothy. *Sex Machines: Photographs and Interviews.* Los Angeles: Daniel 13/Process, 2005.

Frank, Robert T. "The Formation of an Artificial Vagina Without Operation." *American Journal of Obstetrics and Gynecology* 14: 712–718 (1927).

Levins, Hoag. *American Sex Machines: The Hidden History of Sex at the U.S. Patent Office.* Holbrook, Mass.: Adams Media, 1996.

Tiefer, Leonore. "Historical, Scientific, Clinical, and Feminist Criticisms of 'The Human Sexual Response Cycle' Model." *Annual Review of Sex Research* 2: 1–23 (1991).

three | The Princess and Her Pea

Barker-Benfield, Ben. "Sexual Surgery in Late-Nineteenth-Century America." *International Journal of Health Services* 5 (2): 279–298 (1975).

Bertin, Celia. *Marie Bonaparte, A Life.* New York: Harcourt Brace Jovanovich, 1982.

Bonaparte, Marie. "Les Deux Frigidités de la Femme." *Bulletin de la société de sexologie* 1: 161–170 (1932).

Fisher, C. M. "Phantom Erection After Amputation of Penis." *Canadian Journal of Neurological Sciences* 26 (1): 53–56 (1999).

Freud, Sigmund. *New Introductory Lectures on Psychoanalysis.* New York: W. W. Norton, 1965.

Hoch, Zwi. "Vaginal Erotic Sensitivity by Sexological Examination." *Acta Obstetricia et Gynecologica Scandinavica* 65: 767–773 (1986).

Levin, R. J. "Wet and Dry Sex—The Impact of Cultural Influence in Modifying Vaginal Function." *Sexual and Relationship Therapy* 20 (4): 465–474 (2005).

———. "VIP, Vagina, Clitoral and Periurethral Glans—An Update on Human Female Genital Arousal." *Experimental and Clinical Endocrinology* 98 (2): 61–69 (1991).

Lewis, Carolyn Herbst. "Waking Sleeping Beauty: The Premarital Pelvic Exam and Heterosexuality During the Cold War." *Journal of Women's History* 17 (4): 87–110 (2005).

Lloyd, Jillian, et al. "Female Genital Appearance: 'Normality' Unfolds." *British Journal of Obstetrics and Gynaecology* 112: 643–646.

Mondaini, N., et al. "Penile Length Is Normal in Most Men Seeking Penile Lengthening Procedures." *International Journal of Impotence Research* 14: 283–286 (2002).

Narjani, A. E. [Marie Bonaparte]. "Considérations sur les causes anatomiques de la frigidité chez la femme." *Bruxelles-Médical* No. 42, Ap. 27: 768–778 (1924).

Neuhaus, Jessamyn. "The Importance of Being Orgasmic: Sexuality, Gender, and Marital Sex Manuals in the United States, 1920–1963." *Journal of the History of Sexuality* 9 (4): 447–473 (2000).

Velde, Theodoor H. van de. *Ideal Marriage.* New York: Random House, 1930.

Zeigerman, Joseph H., and Jay Gillenwater. "Coitus per Urethram and the Rigid Hymen." *Journal of the American Medical Association* 194 (8): 167–168 (1965).

four | The Upsuck Chronicles

Beck, Joseph. "How Do the Spermatozoa Enter the Uterus?" *American Journal of Obstetrics and Diseases of Women and Children* VII (3): 353–395 (1874).

Fox, C. A., and Beatrice Fox. "A Comparative Study of Coital Physiology, with Special Reference to the Sexual Climax." *Journal of Reproduction and Fertility* 24: 319–336 (1971).

Gallup, Gordon G., Jr., Rebecca L. Burch, and Steven M. Platek. "Does Semen Have Antidepressant Properties?" *Archives of Sexual Behavior* 31 (3): 289–293 (2002).

Goldfoot, D. A., et al. "Behavioral and Physiological Evidence of Sexual Climax in the Female Stump-Tailed Macaque." *Science* 208: 1477–1478 (1980).

Kinsey, Alfred C., Clyde E. Martin, and Wardell B. Pomeroy. *Sexual Behavior in the Human Male.* Philadelphia: W. B. Saunders, 1948.

Kunz, G., et al. "The Dynamics of Rapid Sperm Transport Through the Female Genital Tract: Evidence from Vaginal Sonography of Uterine Peristalsis and Hysterosalpingoscintigraphy." *Human Reproduction* 11 (3): 627–632 (1996).

Levin, Roy J. "The Involvement of the Human Cervix in Reproduction and Sex." *Sexual and Relationship Therapy* 20 (2): 251–260 (2005).

———. "The Physiology of Sexual Arousal in the Human Female: A Recreational and Procreational Synthesis." *Archives of Sexual Behavior* 31 (5): 405–411 (2002).

Madsen, Mads Thor, Johnny Mathiasen, and Dorthe Rønn Olesen. "Effect of Human Stimulation of Sows on Oxytocin in the Blood During Artificial Insemination." National Committee for Pig Production, Report No. 532, January 11, 2001.

Perry, Enos. *The Artificial Insemination of Farm Animals.* New Brunswick: Rutgers University Press, 1952.

Spallanzani, Lazzaro. *Dissertations Relative to the Natural History of Animals and Vegetables.* London: J. Murray, 1789.

Talmey, B. S. "Birth Control and the Physician." *New York Medical Journal.* June 23, 1917, pp. 1185–1192.

VanDemark, N. L., and R. L. Hays. "Uterine Motility Responses to Mating." *American Journal of Physiology* 170: 518–552 (1952).

Yamonaka, Herbert S., and A. L. Soderwall. "Transport of Spermatozoa Through the Female Genital Tract of Hamsters." *Fertility and Sterility* 11: 470–474 (1960).

five | What's Going On in There?

Deng, Jing, et al. "Real-Time Three-Dimensional Ultrasound Visualization of Erection and Artificial Coitus." *International Journal of Andrology* 29: 374–379 (2006).

Foldes, P., and O. Buisson. "Clitoris and G Spot: An Intimate Affair." *Gynécologie obstétrique & fertilité* 35: 3–5 (2007).

Gallup, Jr., Gordon, et al. "The Human Penis as a Semen Displacement Device." *Evolution and Human Behavior* 24 (4): 277–289 (2003).

Meizner, Israel. "Sonographic Observation of in Utero Fetal 'Masturbation.'" *Journal of Ultrasound in Medicine* 6 (2): 111 (1987).

Morris, A. G. "On the Sexual Intercourse Drawings of Leonardo da Vinci." *South African Medical Journal* 69: 510–513 (1986).

Sabelis, Ida. "To Make Love as a Testee." Acceptance speech for 2000 Ig Nobel Prize. www.improbable.com/airchives/paperair/volume7/v7i1/sabel-speech-7-1.html.

Schultz, Willibrord Weijmar, et al. "Magnetic Resonance Imaging of Male and Female Genitals During Coitus and Female Sexual Arousal." *British Medical Journal* 319: 1596–1600 (1999).

six | The Taiwanese Fix and the Penile Pricking Ring

Berardinucci, D., et al. "Surgical Treatment of Penile Veno-Occlusive Dysfunction: Is It Justified?" *Urology* 47 (1): 88–92 (1996).

Chen, Shyh-Chyan, et al. "The Progression of the Penile Vein: Could It Be Recurrent?" *Journal of Andrology* 26 (1): 53–59 (2005).

Darmon, Pierre. *Trial by Impotence: Virility and Marriage in Pre-Revolutionary France*. London: Chatto & Windus, Hogarth Press, 1985.

Howe, Joseph W. *Excessive Venery, Masturbation, and Continence: The Etiology, Pathology, and Treatment of the Diseases Resulting from Venereal Excesses, Masturbation, and Continence*. New York: E. B. Treat, 1896.

Institoris, Heinrich. *Malleus Maleficarum*. New York: Benjamin Blom, 1970 (originally published 1491).

Levin, R. J. "Masturbation and Nocturnal Emissions: Possible Mechanisms for Minimising Teratozoospermie and Hyperspermie in Man." *Medical Hypotheses* 1: 130–131 (1975).

Levins, Hoag. *American Sex Machines: The Hidden History of Sex at the U.S. Patent Office*. Holbrook, Mass.: Adams Media, 1996.

Loe, Meika. *The Rise of Viagra: How the Little Blue Pill Changed Sex in America*. New York: New York University Press, 2004.

Stengers, Jean, and Anne Van Neck. *Masturbation: The History of a Great Terror.* Translated by Kathryn Hoffmann. New York: Palgrave Macmillan, 2001.

Tissot, M. *Onanism; or, A Treatise upon the Disorders Produced by Masturbation.* New York: Garland Publications, 1985 (originally published 1766).

Vale, J. A., et al. "Venous Leak Surgery: Long-Term Follow-Up of Patients Undergoing Excision and Ligation of the Deep Dorsal Vein of the Penis." *British Journal of Urology* 76 (2): 192–195 (1995).

Wooten, Joe S. "Ligation of the Dorsal Vein of the Penis as a Cure for Atonic Impotence." *Texas Medical Journal* XVIII (1): 324–329 (1902).

seven | The Testicle Pushers

Androutsos, Georges. "Skevos Zervos (1875–1966) et les premières greffes testiculaires du singe à l'homme." *Histoires des sciences médicales* 37 (4): 449–456 (2004).

Aristotle (?). *Aristotle's Masterpiece; or, The Secrets of Generation.* London: printed by the booksellers, 1755 (reprinted 1986).

BBC News (online). "Pandas Unexcited by Viagra." September 9, 2002, World: Asia-Pacific.

Dorey, Grace. "Pelvic Floor Muscle Exercises for Erectile Dysfunction." *BJU International* 96 (4): 595–597 (2005).

Gittings, John. "The Demise of the Panda." *Guardian* (London), June 12, 2002.

Hamilton, David. *The Monkey Gland Affair.* London: Chatto & Windus, 1986.

Highley, Keith. "The Market for Tiger Products on Taiwan: A Survey." Earthtrust Taiwan report, March 1993.

Lee, R. Alton. *The Bizarre Careers of John R. Brinkley.* Lexington: University Press of Kentucky, 2002.

Lin, J. H., W. W. Chen, and L. S. Wu. "Acupuncture Treatments for Animal Reproductive Disorders." *Web-Journal of Acupuncture.* (Not dated or numbered; use site search engine.)

Lydston, G. Frank. "Sex Gland Implantation." *Journal of the American Medical Association* 66 (20): 1540–1543 (1916).

———. "Further Observations on Sex Gland Implantation." *Journal of the American Medical Association* 72 (6): 396–398 (1919).

Read, Bernard E. *Chinese Materia Medica: Animal Drugs.* From the *Pen Ts'ao Kang Mu* by Li Shih-chen, 1597. Taipei: Southern Materials Center, 1976.

Stanley, L. L. "An Analysis of One Thousand Testicular Substance Implantations." *Endocrinology* 6: 787–794 (1922).

Sun (Malaysia). "Viagra No Joy for These Women." March 21, 1999, p. A36.

Voronoff, Serge. *Rejuvenation by Grafting.* New York: Adelphi Co., 1925.

eight | Re-Member Me

Bhanganada, Kasian, et al. "Surgical Management of an Epidemic of Penile Amputations in Siam." *American Journal of Surgery* 146: 376–382 (1983).

Carson, Culley C. "Penile Prosthesis Implantation: Surgical Implants in the Era of Oral Medication." *Urological Clinics of North America* 32: 503–509 (2005).

Chappell, Buford S. "Relief of Impotency by Cartilage Implants: Presentation of a Technic." *Journal of the South Carolina Medical Association,* February 1952, pp. 31–34.

Dunn, M. E., and J. E. Trost. "Male Multiple Orgasms: A Descriptive Study." *Archives of Sexual Behavior* 18 (5): 377–387 (1989).

Fisher, C. M. "Phantom Erection After Amputation of Penis: Case Description and Review of Relevant Literature on Phantoms." *Canadian Journal of Neurological Sciences* 26: 53–56 (1999).

Gray, P., and B. Campbell. "Erectile Dysfunction and Its Correlates Among the Ariaal of Northern Kenya." *International Journal of Impotence Research* 17: 445–449 (2005).

McLaren, Robert H., and David M. Barrett. "Patient and Partner Satisfaction with the AMS 700 Penile Prosthesis." *Journal of Urology* 147: 62–65 (1992).

Mishra, Baikunthnath, and Nilamadhab Kar. "Genital Self Amputation for Urinary Symptom Relief." *Indian Journal of Psychiatry* 43 (4): 342–344 (2001).

Salama, N. "Satisfaction with the Malleable Penile Prosthesis Among

Couples from the Middle East: Is It Different from That Reported Elsewhere?" *International Journal of Impotence Research* 16: 175–180 (2004).

Waugh, A. C. "Autocastration and Biblical Delusions in Schizophrenia." *British Journal of Psychiatry* 149: 656–659 (1986).

nine | The Lady's Boner

Karacan, Ismet, Constance Moore, and Sezai Sahmay. "Measurement of Pressure Necessary for Vaginal Penetration." *Sleep Research* 14: 269 (1985).

Karacan, I., A. L. Rosenbloom, and R. L. Williams. "The Clitoral Erection Cycle During Sleep." *Psychophysiology* 7: 338 (abstract), (1970).

Kolata, Gina. "Pfizer Gives Up Testing Viagra on Women." *New York Times*, Science Times/Health, February 28, 2004.

Laan, Ellen, et al. "The Enhancement of Vaginal Vasocongestion by Sildenafil in Healthy Premenopausal Women." *Journal of Women's Health and Gender-Based Medicine* 11 (4): 357–385 (2002).

Loe, Meika. *The Rise of Viagra: How the Little Blue Pill Changed Sex in America.* New York: New York University Press, 2004.

Maravilla, Ken, et al. "Dynamic MR Imaging of the Sexual Arousal Response in Women." *Journal of Sex and Marital Therapy* 29 (s): 71–76 (2003).

Park, K., et al. "Vasculogenic Female Sexual Dysfunction: The Hemodynamic Basis for Vaginal Engorgement Insufficiency and Clitoral Erectile Insufficiency." *International Journal of Impotence* 9: 27–37 (1997).

ten | The Prescription-Strength Vibrator

Aetios. *Aetios of Amida: The Gynaecology and Obstetrics of the VIth Century, A.D.* Translated and annotated by James B. Ricci. Philadelphia: Blakiston, 1950.

Bohlen, Joseph G., et al. "Heart Rate, Rate-Pressure Product, and Oxygen Uptake During Four Sexual Activities." *Archives of Internal Medicine* 144: 1745–1748 (1984).

Brindley, Giles. "Electroejaculation: Its Technique, Neurological Implications, and Uses." *Journal of Neurology, Neurosurgery, and Psychiatry* 44: 9–18 (1981).

Butt, Dorcus S. "The Sexual Response as Exercise: A Brief Review and Theoretical Proposal." *Sports Medicine* 9 (6): 330–343.

Davey Smith, G., S. Frankel, and J. Yarnell. "Sex and Death: Are They Related? Findings from the Caerphilly Cohort Study." *BMJ* (Clinical Research ed.): 315: 1641–1644 (1991).

Eccles, Audrey. *Obstetrics and Gynaecology in Tudor and Stuart England.* Kent, Ohio: Kent State University Press, 1982.

Horne, Herbert H., David P. Paul, and Donald Munro. "Fertility Studies in the Human Male with Traumatic Injuries of the Spinal Cord and Cauda Equina." *New England Journal of Medicine* 239 (25): 959–961 (1948).

Imami, Riazul H., and Miftah Kemal. "Vacuum Cleaner Use in Autoerotic Death." *American Journal of Forensic Medicine and Pathology* 9 (3): 246–248 (1988).

Klotz, Laurence. "How (Not) to Communicate New Scientific Information: A Memoir of the Famous Brindley Lecture." *BJU International* 96 (7): 956–957 (2005).

Maines, Rachel P. *The Technology of Orgasm: "Hysteria," the Vibrator, and Women's Sexual Satisfaction.* Baltimore: Johns Hopkins University Press, 1999.

Peleg, Roni, and Aya Peleg. "Case Report: Sexual Intercourse as Potential Treatment for Intractable Hiccups." *Canadian Family Physician* 46: 1631–1632 (2000).

Schroder, Maryann. "A Clinical Trial of the Eros Therapy for the Treatment of Sexual Dysfunction in Postmenopausal Women." Paper presented at March 2002 annual meeting of the Society for Sex Therapy and Research.

Soranus. *Soranus' Gynecology.* Translated and annotated by Owsei Temkin. Baltimore: Johns Hopkins University Press, 1991.

Brackett, Nancy L., et al. "An Analysis of 653 Trials of Penile Vibratory Stimulation in Men with Spinal Cord Injury." *Journal of Urology* 159 (6): 1931–1934 (1998).

Chuang, Yao-Chung, et al. "Tooth-Brushing Epilepsy with Ictal Orgasms." *Seizure* 13: 179–182 (2004).

Dosemeci, L., et al. "Frequency of Spinal Reflex Movements in Brain-Dead Patients." *Transplantation Proceedings* 36 (1): 17–19 (2004).

Komisaruk, Barry R., Carlos Beyer-Flores, and Beverly Whipple. *The Science of Orgasm.* Baltimore: Johns Hopkins University Press, 2006.

Komisaruk, Barry R., et al. "Brain Activation During Vaginocervical Self-Stimulation and Orgasm in Women with Complete Spinal Cord Injury: fMRI Evidence of Mediation by the Vagus Nerves." *Brain Research* 1024: 77–88 (2004).

Levin, Roy, and Gorm Wagner. "Orgasm in Women in the Laboratory —Qualitative Studies on Duration, Intensity, Latency, and Vaginal Blood Flow." *Archives of Sexual Behavior* 14 (5): 439–449.

Sexuality Reborn: Sexuality Following Spinal Cord Injury. Videocassette. Produced by Marca L. Sipski, Craig Alexander, and Mary Eyles. The Kessler Rehabilitation Research and Education Corp., 1993.

Sipski, Marca L., and Craig J. Alexander. *Sexual Function in People with Disability and Chronic Illness.* Gaithersburg, MD: Aspen Publishers, 1997.

———, and Raymond Rosen. "Sexual Arousal and Orgasm in Women: Effects of Spinal Cord Injury." *Annals of Neurology* 49 (1): 35–44.

———. "Spinal Cord Injuries and Orgasm: A Review." Submitted for publication.

Van Der Schoot, D. K. E., and A. F. G. V. M. Ypma. "Seminal Vesiculectomy to Resolve Defecation-Induced Orgasm." *BJU International* 90: 761–762 (2002).

Whipple, Beverly, Barry R. Komisaruk, and Gina Ogden. "Physiological Correlates of Imagery-Induced Orgasm in Women." *Archives of Sexual Behavior* 21 (2): 121–133 (1992).

Women's Sexuality After SCI: Understanding the Changes and Finding

New Ways to Respond. Videocassette, 18 minutes. Marca L. Sipski & University of Miami School of Medicine, 2003.

twelve | Mind over Vagina

Chivers, Meredith L., et al. "A Sex Difference in the Specificity of Sexual Arousal." *Psychological Science* 15 (11): 736–744 (2004).

Dove, Natalie L., and Micheal W. Wiederman. "Cognitive Distraction and Women's Sexual Functioning." *Journal of Sex and Marital Therapy* 26: 67–78 (2000).

Down, J. Langdon H. "Influence of the Sewing Machine on Female Health." *British Medical Journal*, January 12, 1867, pp. 26–27.

Hamilton, Lisa Dawn, and Cindy M. Meston. "The Effect of Sexual Activity on Testosterone in Women." Presentation at the 2006 meeting of the International Society for the Study of Women's Sexual Health.

Janssen, Erik, Deanna Carpenter, and Cynthia A. Graham. "Selecting Films for Sex Research: Gender Differences in Erotic Film Preference." *Archives of Sexual Behavior* 32 (3): 243–251 (2003).

Levin, Roy J., and Willy van Berlo. "Sexual Arousal and Orgasm in Subjects Who Experience Forced or Non-consensual Sexual Stimulation: A Review." *Journal of Clinical Forensic Medicine* 11: 82–88 (2004).

Rachman, S., and R. J. Hodgson. "Experimentally Induced 'Sexual Fetishism': Replication and Development." *Psychological Record* 18: 25–27 (1968).

Rellini, Alessandra H., et al. "The Relationship Between Women's Subjective and Physiological Sexual Arousal." *Psychophysiology* 42: 116–124 (2005).

Sawatsky, John. *Men in the Shadows: The RCMP Security Service.* Toronto: Doubleday Canada, 1980.

thirteen | What Would Allah Say?

Busch, David, and James R. Starling. "Rectal Foreign Bodies: Case Reports and a Comprehensive Review of the World's Literature." *Surgery* 100 (3): 513–519.

Levin, R. J. "Do Women Gain Anything from Coitus Apart from Pregnancy? Changes in the Human Female Genital Tract Activated by Coitus." *Journal of Sex and Marital Therapy* 29 (s): 59–69 (2003).

Shafik, Ahmed. "Effect of Different Types of Textiles on Sexual Activity." *European Urology* 24: 375–380 (1993).

———. "The Peno-Motor Reflex: Study of the Response of the Puborectalis and Levator Ani Muscles to Glans Penis Stimulation." *International Journal of Impotence Research* 7: 239–246 (1995).

———. "Vaginocavernosus Reflex: Clinical Significance and Role in the Sexual Act." *Gynecologic and Obstetric Investigation* 35: 114–117 (1993).

———. "Vagino-Levator Reflex: Description of a Reflex and Its Role in Sexual Performance." *European Journal of Obstetrics and Gynecology* 60: 161–164 (1995).

———, et al. "Flaturia: Passage of Flatus at Coitus. Incidence and Pathogenesis." *Archives of Gynecology and Obstetrics,* August 16, 2006 (epub).

fourteen | Monkey Do

Cutler, Winnifred B., Erika Friedmann, and Norma L. McCoy. "Pheromonal Influences on Sociosexual Behavior in Men." *Archives of Sexual Behavior* 27 (1): 1–13 (1998).

Doty, Richard L., et al. "Changes in the Intensity and Pleasantness of Human Vaginal Odors During the Menstrual Cycle." *Science* 190: 1316–1317.

Goldfoot, D. A., et al. "Lack of Effect of Vaginal Lavages and Aliphatic Acids on Ejaculatory Responses in Rhesus Monkeys: Behavioral and Chemical Analyses." *Hormones and Behavior* 7: 1–27 (1976).

Kirk-Smith, M. D., and D. A. Booth. "Effect of Androstenone on Choice of Location in Others' Presence." In *Proceedings of the Seventh International Symposium on Olfaction and Taste.* London and Washington: IRL Press, 1980.

Levin, Roy J. "Smells and Tastes: Their Putative Influence on Sexual Activity in Humans." *Sexual and Relationship Therapy* 19 (4): 451–462 (2004).

Martin, David E., and Kenneth G. Gould. "The Male Ape Genital Tract and Its Secretions." In *Reproductive Biology of the Great Apes*. Edited by Charles E. Graham. New York: Academic Press, 1981.

Michael, R. P., and E. B. Keverne. "Pheromones in the Communication of Sexual Status in Primates." *Nature* 218: 746–749 (1968).

Morris, Naomi M., and J. Richard Udry. "Pheromonal Influences on Human Sexual Behavior: An Experimental Search." *Journal of Biosocial Science* 10: 147–157 (1978).

Simon, J., et al. "Testosterone Patch Increases Sexual Activity and Desire in Surgically Menopausal Women with Hypoactive Sexual Desire Disorder." *Journal of Clinical Endocrinology and Metabolism* 90 (9): 5226–5233 (2005).

Tutin, Caroline E. G., and Patrick R. McGinnis. "Chimpanzee Reproduction in the Wild." In *Reproductive Biology of the Great Apes*. Edited by Charles E. Graham. New York: Academic Press, 1981.

Wallen, Kim. "Desire and Ability: Hormones and the Regulation of Female Sexual Behavior." *Neuroscience and Biobehavioral Reviews* 14: 233–241 (1990).

———. "Risky Business: Social Context and Hormonal Modulation of Primate Sexual Desire." In *Reproduction in Context*, edited by Kim Wallen and J. E. Schneider. Cambridge: MIT Press, 2000.

———. "Sex and Context: Hormones and Primate Sexual Motivation." *Hormones and Behavior* 40: 339–357 (2001).

Wysocki, Charles J., and George Preti. "Facts, Fallacies, Fears, and Frustrations with Human Pheromones." *Anatomical Record Part A: Discoveries in Molecular, Cellular, and Evolutionary Biology* 28 (1): 1201–1211 (2004).

fifteen | "Persons Studied in Pairs"

Masters, William H., and Virginia Johnson. *Homosexuality in Perspective*. Boston: Little, Brown, 1979.

BONK

Mary Roach

BONK

Mary Roach

DISCUSSION QUESTIONS

1. Why might Mary Roach have chosen to make herself and her husband human subjects in lab-based studies of sex?

2. How does humor help Roach tackle the myriad questions surrounding human sex lives and practices?

3. Mary Roach writes, "Sex is far more than the sum of its moving parts" (p. 128). Unpack that statement. What insight does it provide into the functions and limitations of lab-based, physiological studies of sex?

4. Freudian theory holds that grown women who rely on the clitoris for sexual gratification are stuck in a childlike state. According to the theory, Roach says, "This 'phallic' phase is supposed to end at puberty, when a woman embraces her proper role as a passive, feminine being." Freud himself wrote that "the clitoris should . . . hand over its sensitivity, and at the same time its importance, to the vagina" (p. 81). Roach presents research that subverts Freud's theory about the separation of the clitoris and vagina. How has physiological science offered a defense against Freud's theories on female sexuality? Why does this matter?

5. Roach notes that the linking of sexual delight and fertility dates as far back as Western medicine itself. Does this idea—no orgasm, no babies—surprise you? How have ideas about fertility shaped our understanding of sexual gratification?

6. The nineteenth-century physician Joseph Beck felt confident that some sort of uterine "upsuck" occurred during a female orgasm— "upsuck" that could pull sperm toward an egg for fertilization. But sex physiologist Roy Levin points out that "sperm straight out of

the penis are not yet up to the job of fertilizing an egg. They need time to capacitate" (p. 106). What is the lesson here in regard to fertility science?

7. Roach describes the introduction of Viagra to consumers. "In 1998," she writes, "Pfizer—with a cadre of media-savvy urologists in tow—launched a massive publicity campaign to announce an exciting new approach to impotence. Only it wasn't called impotence anymore; it was 'erectile dysfunction'" (p. 141). Why do you think the language changed? Does one terminology sound more "medical" than the other? Why might that be significant?

8. "*Homo sapiens*," Roach writes, "is one of the few species on earth that care if they're seen having sex" (p. 279). How do you react to this idea? What insight might this provide into the biological pressures and cultural forces at work in our sex lives?

9. Like *Six Feet Over* and *Stiff*, *Bonk* involves a wide-ranging tour of the human body. How would you compare these books? What kind of research techniques and writing style do all three books employ?

10. Roach remarks that "The media's ubiquitous coverage of sex and sex research . . . have chipped away at the taboos that kept couples from talking openly with each other about the sex they were having" (p. 302). Do you agree or disagree? Has journalism made sex easier to discuss? And has sex become a more admissible subject of scientific research?

Don't miss other titles by best-selling author

Mary Roach

MARYROACH.NET

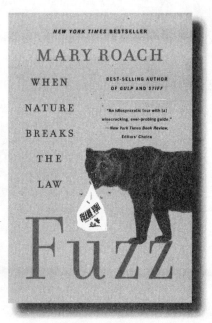

NEW YORK TIMES BESTSELLER

#1 LOS ANGELES TIMES BESTSELLER

#1 INDIE HARDCOVER NONFICTION BESTSELLER

A PUBLISHERS WEEKLY BEST NONFICTION BOOK OF 2021

LONGLISTED FOR THE 2022 ANDREW CARNEGIE MEDAL FOR
EXCELLENCE IN NONFICTION

"Powerfully propelled by the force of Roach's unflinching fascination with the weird, the gross and the downright improbable."　　　—Amelia Urry, *Washington Post*

NEW YORK TIMES BESTSELLER

"Her funniest and most sparkling book."

—Janet Maslin, *New York Times*

NEW YORK TIMES BESTSELLER

"A mirthful, informative peek behind the curtain of military science." —*Washington Post*

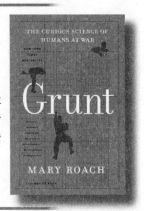

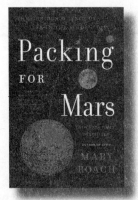

NEW YORK TIMES BESTSELLER

"An often hilarious, sometimes queasy-making catalog of the strange stuff devised to permit people to survive in an environment for which their bodies are stupendously unsuited."

—M. G. Lord, *New York Times Book Review*

W. W. NORTON & COMPANY
Independent Publishers Since 1923